SOLDIER'S MANUAL AND TRAINER'S GUIDE
MOS 18E

STP 31-18E34-SM-TG

Special Forces Communications Sergeant
Skill Levels 3 and 4

February 2010

DISTRIBUTION RESTRICTION:
Distribution authorized to U.S. Government agencies only to protect technical or operational information from automatic dissemination under the International Exchange Program or by other means. This determination was made 11 December 2009. Other requests for this document must be referred to Commander, U.S. Army John F. Kennedy Special Warfare Center and School, ATTN: AOJK-DTD-SF, Fort Bragg, NC 28310-9610.

FOREIGN DISCLOSURE:
This publication has been reviewed by the product developers in coordination with the U.S. Army John F. Kennedy Special Warfare Center and School foreign disclosure authority. This product is releasable to students from foreign countries on a case-by-case basis.

DESTRUCTION NOTICE:
Destroy by any method that will prevent disclosure of contents or reconstruction of the document.

**HEADQUARTERS
DEPARTMENT OF THE ARMY**

This publication is available at
Army Knowledge Online (www.us.army.mil) and
General Dennis J. Reimer Training and Doctrine
Digital Library at (www.train.army.mil).

STP 31-18E34-SM-TG

¹SOLDIER TRAINING
PUBLICATION
No. 31-18E34-SM-TG

HEADQUARTERS
DEPARTMENT OF THE ARMY
Washington, DC, 8 Febuary 2010

SOLDIER'S MANUAL and TRAINER'S GUIDE
MOS 18E
Skill Levels 3 and 4

TABLE OF CONTENTS

PAGE

TABLE OF CONTENTS i
PREFACE iv

CHAPTER 1   INTRODUCTION 1-1
    1-1.   General 1-1
    1-2.   Soldier's Responsibilities 1-1
    1-3.   NCO Self-Development and the Soldier's Manual 1-2
    1-4.   Training Support 1-2

CHAPTER 2   TRAINER'S GUIDE 2-1
    2-1.   General 2-1
    2-2.   Subject Area Codes 2-2
    2-3.   Duty Position Training Requirements 2-2
    2-4.   Critical Tasks List 2-3

CHAPTER 3   MOS/SKILL LEVEL TASKS 3-1

Skill Level 3
Subject Area 1: Computer Applications

113-580-1035   Install a Tactical Local Area Network 3-1
113-581-1001   Operate a Personal Computer 3-3
331-18E-3075   Transmit a Digital Image 3-4

DISTRIBUTION RESTRICTION: Distribution authorized to U.S. Government agencies only to protect technical or operational information from automatic dissemination under the International Exchange Program or by other means. This determination was made 11 December 2009. Other requests for this document must be referred to Commander, U.S. Army John F. Kennedy Special Warfare Center and School, ATTN: AOJK-DTD-SF, Fort Bragg, NC 28310-9610.

FOREIGN DISCLOSURE: This publication has been reviewed by the product developers in coordination with the U.S. Army John F. Kennedy Special Warfare Center and School foreign disclosure authority. This product is releasable to students from foreign countries on a case-by-case basis.

DESTRUCTION NOTICE: Destroy by any method that will prevent disclosure of contents or reconstruction of the document.

¹This publication supersedes STP 31-18E34-SM-TG, 5 March 2007.

STP 31-18E34-SM-TG

Subject Area 2: Communication Procedures

113-571-1003	Communicate in a Radio Net	3-5
113-571-1004	Operate in Radio Nets	3-7
113-573-0001	Check Signal Security Procedures	3-8
113-573-1004	Conduct Communications Security Inspections	3-10
113-573-2029	Conduct Shift-to-Shift Inventory of Communications Security Material	3-11
113-573-6001	Recognize Electronic Attack and Implement Electronic Protection	3-13
113-573-8006	Use an Automated Signal Operation Instruction	3-16
113-573-9012	Destroy Classified Material	3-22
113-611-6112	Prepare Signal Annex of the Operation Order	3-25
158-350-0001	Implement Defense Information Operations	3-27
331-18E-3032	Develop a Communications Plan	3-29
331-18E-3058	Occupy a Transmission Site	3-30
331-18E-3076	Compute Communications Equipment Electrical Requirements	3-32
331-18E-3077	Manage Classified Material	3-33

Subject Area 3: Communication Systems

113-587-1064	Prepare Single-Channel Ground and Airborne Radio System (Manpack) for Operation	3-34
113-587-1067	Install Single-Channel Ground and Airborne Radio Systems ICOM With or Without the AN/VIC-1 or AN/VIC-3	3-35
113-596-1052	Construct Vertical Half-Rhombic Antenna	3-37
113-596-1056	Construct a Long-Wire Antenna	3-38
113-596-1070	Construct a Doublet Antenna	3-39
113-609-2006	Operate Simple Key Loader AN/PYQ-10	3-40
113-609-2053	Operate Automated Net Control Device AN/CYZ-10	3-43
113-609-4000	Restore the Simple Key Loader AN/PYQ-10	3-45
113-620-1028	Install Radio Set AN/GRC-193A or Similar Radio Sets	3-47
113-620-1040	Install Improved High Frequency Radio Set AN/GRC-213 or a Similar System	3-49
331-18E-3019	Employ International Maritime Satellite Terminal	3-51
331-18E-3024	Operate the AN/PSC-5 in Satellite Communications Mode	3-54
331-18E-3025	Operate the AN/PSC-5C/D in Demand Assignment Multiple Access Mode	3-56
331-18E-3027	Update Crypto Fill Keys on the AN/PSC-5	3-60
331-18E-3028	Perform Unit-Level Preventative Maintenance Checks and Services on Communications Equipment	3-61
331-18E-3029	Use a Multimeter to Perform a Continuity Check and Voltage Check	3-62
331-18E-3042	Construct a Clandestine Antenna	3-63
331-18E-3043	Install Antenna Group OE-452/PRC	3-67
331-18E-3044	Construct Sloping-Vee Antenna	3-70
331-18E-3047	Construct Expedient RC-292 Antenna	3-73
331-18E-3048	Employ Radio Set AN/PRC-137F	3-76
331-18E-3052	Employ Radio Set AN/PSC-5D	3-79
331-18E-3056	Employ Radio Set AN/PRC-148	3-84
331-18E-3070	Employ Iridium Emergency Communications	3-89
331-18E-3073	Construct a Slant-Wire Antenna	3-91

Subject Area 4: Communications Planning

113-611-6002	Plan Frequency Modulated Voice and Data Communications Net	3-93
331-18E-3081	Plan for the Employment of Tactical Radio Systems	3-95

331-18E-3083 Install Mission Planning Kit 3-96
Skill Level 4
Subject Area 2: Communication Procedures
331-18E-4003 Establish a Special Operations Task Force Signal Center 3-98
Subject Area 4: Communications Planning
331-18E-4004 Coordinate Signal Activities With Other Units 3-101
331-18E-4005 Supervise Signal Augmentation 3-102
GLOSSARY **Glossary-1**
REFERENCES **References-1**

STP 31-18E34-SM-TG

Preface

Soldier training publication (STP) 31-18E34-SM-TG, *Special Forces Communications Sergeant, Skill Levels 3 and 4*, contains standardized training objectives (in the form of task summaries) that can be used to train and evaluate Soldiers on critical tasks that support unit missions during wartime and peacetime operations.

Purpose

Commanders, trainers, and Soldiers use this manual with the STP 21-1-SMCT, *Soldier's Manual of Common Tasks, Warrior Skills, Level 1*; STP 21-24-SMCT, *Soldier's Manual of Common Tasks, Warrior Leader, Skill Levels 2, 3, and 4*; and Army Training and Evaluation Programs (ARTEPs) to establish effective training plans and programs that integrate Soldier, leader, and collective tasks.

Scope

Noncommissioned officers (NCOs) holding military occupational specialty (MOS) 18E, skill levels (SLs) 3 and 4, will have access to this STP. Trainers and first-line supervisors will ensure it is available in the Soldiers' work area, unit learning center, and unit libraries. However, there is no requirement for each Soldier to be provided an individual copy. Commanders will ensure this STP is readily available to all Soldiers.

Applicability

All tasks in this STP are applicable to both the Active Army and Army National Guard (ARNG) Component Soldiers. However, due to differences in tables of organization and equipment (TOEs) and missions, some tasks may not apply to all Special Forces (SF) units.

Administrative Information

The proponent of this STP is the United States Army John F. Kennedy Special Warfare Center and School (USAJFKSWCS). The users of this STP are encouraged to recommend changes and submit comments for its improvement. Key comments to specific page, paragraph, and line of the text in which the change is recommended. Provide reasons for each comment to ensure understanding and complete evaluation.

Prepare comments on Department of the Army (DA) Form 2028 and forward them to Commander, United States Army John F. Kennedy Special Warfare Center and School (USAJFKSWCS), ATTN: AOJK DTD-SF, Fort Bragg, North Carolina 28310-9610. Unless this STP states otherwise, masculine nouns and pronouns do not refer exclusively to men.

STP 31-18E34-SM-TG

Chapter 1

Introduction

1-1. GENERAL

a. This Soldier's manual (SM) identifies the individual military MOS training requirements for Soldiers in MOS 18E. Commanders, trainers, and Soldiers will use it to plan, conduct, and evaluate individual training in units. This manual is the primary MOS reference to support the self-development and training of the Soldier.

b. Commanders, trainers, and Soldiers use this manual with the STP 21-1-SMCT, STP 21-24-SMCT, and ARTEPs to establish effective training plans and programs that integrate Soldier, leader, and collective tasks.

c. The Army's mission is to mobilize and deploy units trained to accomplish wartime missions. Successful mission accomplishment requires emphasis on individual training. Individual training must focus on performance under the conditions and to the standards expected in wartime. This STP, in conjunction with STP 31-18-SM-TG, *Soldier's Manual and Trainer's Guide: CMF 18, Special Forces Common Skills, Skill Levels 3 and 4*, identify the individual MOS training requirements for Soldiers in MOS 18E, SLs 3 and 4. It is designed to be used by commanders, trainers, and Soldiers to plan, conduct, and evaluate individual training in units.

d. Task summaries outline the wartime performance requirements of each critical task in the SM. They give the Soldier and the trainer the information necessary to prepare, conduct, and evaluate critical task training. As a minimum, task summaries include information the Soldier must know and the skills he must perform to standard for each task. These summaries are, in effect, standardized training objectives that ensure Soldiers do not have to relearn a task upon assignment to a new unit.

e. This publication contains repetitive feedback statements in all its task summaries that read as follows: Score the Soldier GO if all performance measures are passed. Score the Soldier NO-GO if any performance measure is failed. If the Soldier fails any performance measure, show what was done wrong and how to do it correctly.

f. Critical tasks are those essential to successful individual skill performance and survival on the battlefield. Critical tasks for MOS 18E are in Chapter 3 of this publication.

g. Additionally, some task summaries include safety statements and notes. Safety statements (danger, warning, and caution) alert users to the possibility of immediate death, personal injury, or damage to equipment. Notes are short extra supportive explanations relevant to the performance measures.

1-2. SOLDIER'S RESPONSIBILITIES

Each Soldier is responsible for performing individual tasks the first-line supervisor identifies based on the unit's mission-essential task list (METL). The Soldier must perform the task to the standards listed in the SM. If a Soldier has a question about performing a task, or which task in this manual he must perform, he must ask the first-line supervisor for clarification. The first-line supervisor knows how to perform each task or can direct the Soldier to the appropriate training materials.

1-3. NCO SELF-DEVELOPMENT AND THE SOLDIER'S MANUAL

a. Self-development is one of the key components of the leader development program. It is a planned, progressive, and sequential program followed by leaders to enhance and sustain their military competencies. It consists of individual study, research, professional reading, practice, and self-assessment. Under the self-development concept, the NCO, as an Army professional, is responsible for remaining current in all phases of his MOS. The SM is the NCO's primary source in maintaining MOS proficiency.

b. Another important resource for NCO self-development is the Army Correspondence Course Program (ACCP). For information on enrolling in this program and for a list of courses, log on to the ACCP at http://www.atsc.army.mil/accp/AIPDnew.asp.

c. Unit learning centers are also valuable resources for planning self-development programs. They can help access enlisted career maps, training support products, and extension training materials.

1-4. TRAINING SUPPORT

This manual includes the following additional training support information:

a. Glossary. The glossary is a single comprehensive list of acronyms, abbreviations, and definitions.

b. References. The references section contains two parts—required and related. Required references are necessary for the Soldier to do the task. These references are listed in the conditions statement and at the end of the task summary. Related references are materials that provide more detailed information and a more thorough explanation of task performance. All references are listed at the end of the task summary.

STP 31-18E34-SM-TG

Chapter 2

Trainer's Guide

2-1. GENERAL

The MOS Training Plan (MTP) identifies the essential components of a unit training plan for individual training. Units have different training needs and requirements based on differences in environment, location, equipment, dispersion, and similar factors. Therefore, the MTP should be used as a guide for conducting unit training and not a rigid standard. The MTP consists of two parts. Each part is designed to assist the commander in preparing a unit training plan which satisfies integration, cross training, training up, and sustainment training requirements for Soldiers in this MOS.

Part One of the MTP shows the relationship of an MOS skill level between duty position and critical tasks. These critical tasks are grouped by task commonality into subject areas.

Section I lists subject area numbers and titles used throughout the MTP. These subject areas are used to define the training requirements for each duty position within an MOS.

Section II identifies the total training requirement for each duty position within an MOS and provides a recommendation for cross training and train-up/merger training.

- **Duty Position Column**. This column lists the duty positions of the MOS, by skill level, which have different training requirements.
- **Subject Area Column**. This column lists, by numerical key (see Section I), the subject areas a Soldier must be proficient in to perform in that duty position.
- **Cross-Train Column**. This column lists the recommended duty position for which Soldiers should be cross-trained.
- **Train-Up/Merger Column**. This column lists the corresponding duty position for the next higher skill level or MOSC the Soldier will merge into on promotion.

Part Two lists, by general subject areas, the critical tasks to be trained in an MOS and the type of training required (resident, integration, or sustainment).

- **Subject Area Column**. This column lists the subject area number and title in the same order as Section I, Part One of the MTP.
- **Task Number Column**. This column lists the task numbers for all tasks included in the subject area.
- **Title Column**. This column lists the task title for each task in the subject area.
- **Training Location Column**. This column identifies the training location where the task is first trained to Soldier training publications standards. If the task is first trained to standard in the unit, the word "Unit" will be in this column. If the task is first trained to standard in the training base, it will identify, by brevity code (ANCOC, BNCOC, etc.), the resident course where the task was taught. Figure 2-1 contains a list of training locations and their corresponding brevity codes.

STP 31-18E34-SM-TG

SFQC	Special Forces Qualification Course
AIT	Advanced Individual Training
UNIT	Trained in the Unit
ANCOC	Advanced NCO Course
OSUT	One Station Unit Training
PLDC	Primary Leadership Development Course

Figure 2-1. Training Locations

- **Sustainment Training Frequency Column.** This column indicates the recommended frequency at which the tasks should be trained to ensure Soldiers maintain task proficiency. Figure 2-2 identifies the frequency codes used in this column.

BA	-	Biannually
AN	-	Annually
SA	-	Semiannually
QT	-	Quarterly
MO	-	Monthly
BW	-	Biweekly
WK	-	Weekly

Figure 2-2. Sustainment Training Frequency Codes

- **Sustainment Training Skill Level Column.** This column lists the skill levels of the MOS for which Soldiers must receive sustainment training to ensure they maintain proficiency to Soldier's manual standards.

2-2. SUBJECT AREA CODES

Skill Level 3
1. Computer Applications
2. Communication Procedures
3. Communication Systems
4. Communications Planning

Skill Level 4
2. Communication Procedures
4. Communications Planning

2-3. DUTY POSITION TRAINING REQUIREMENTS

2-4. CRITICAL TASKS LIST

MOS TRAINING PLAN
18E34

CRITICAL TASKS

Task Number	Title	Training Location	Sust Tng Freq	Sust Tng SL
Skill Level 3				
Subject Area 1. Computer Applications				
113-580-1035	Install a Tactical Local Area Network	SFQC	AN	3-4
113-581-1001	Operate A Personal Computer	SFQC	AN	3
331-18E-3075	Transmit a Digital Image	SFQC	SA	3
Subject Area 2. Communication Procedures				
113-571-1003	Communicate in a Radio Net	SFQC	AN	3
113-571-1004	Operate in Radio Nets	SFQC	AN	3-4
113-573-0001	Check Signal Security Procedures	SFQC	AN	3-4
113-573-1004	Conduct Communications Security Inspections	SFQC	AN	3-4
113-573-2029	Conduct Shift-to-Shift Inventory of COMSEC Material	SFQC	AN	3-4
113-573-6001	Recognize Electronic Attack and Implement Electronic Protection	SFQC	AN	3-4
113-573-8006	Use an Automated Signal Operation Instruction	SFQC	AN	3-4
113-573-9012	Destroy Classified Material	SFQC	SA	3
113-611-6112	Prepare Signal Annex of the OPORD	SFQC	AN	3
158-350-0001	Implement Defense Information Operations	SFQC	AN	3-4
331-18E-3032	Develop a Communications Plan	SFQC	SA	3
331-18E-3058	Occupy a Transmission Site	SFQC	SA	3
331-18E-3076	Compute Communications Equipment Electrical Requirements	SFQC	SA	3
331-18E-3077	Manage Classified Material	SFQC	SA	3
Subject Area 3. Communication Systems				
113-587-1064	Prepare Single-Channel Ground and Airborne Radio Systems (Manpack) for Operation	SFQC	AN	3-4
113-587-1067	Install Single-Channel Ground and Airborne Radio Systems ICOM With or Without the AN/VIC-1 or AN/VIC-3	SFQC	AN	3-4
113-596-1052	Construct Vertical Half-Rhombic Antenna	SFQC	SA	1-3
113-596-1056	Construct a Long-Wire Antenna	SFQC	SA	1-3
113-596-1070	Construct a Doublet Antenna	SFQC	SA	1-3
113-609-2006	Operate Simple Key Loader AN/PYQ-10	SFQC	AN	3
113-609-2053	Operate Automated Net Control Device AN/CYZ-10	SFQC	SA	3/4
113-609-4000	Restore the Simple Key Loader AN/PYQ-10	SFQC	QT	3-4
113-620-1028	Install Radio Set AN/GRC-193A or Similar Radio Sets	SFQC	AN	3-4
113-620-1040	Install Improved High Frequency Radio Set AN/GRC-213 or a Similar System	SFQC	AN	3-4

STP 31-18E34-SM-TG

331-18E-3019	Employ International Maritime Satellite Terminal	SFQC	AN	3-4
331-18E-3024	Operate the AN/PSC-5 in Satellite Communications Mode	SFQC	SA	3
331-18E-3025	Operate the AN/PSC-5C/D in Demand Assignment Multiple Access Mode	SFQC	SA	3
331-18E-3027	Update Crypto Fill Keys on the AN/PSC-5	SFQC	SA	3
331-18E-3028	Perform Unit-Level Preventative Maintenance Checks and Services on Communications Equipment	SFQC	SA	3
331-18E-3029	Use a Multimeter to Perform a Continuity Check and Voltage Check	SFQC	SA	3
331-18E-3042	Construct a Clandestine Antenna	SFQC	AN	3-4
331-18E-3043	Install Antenna Group OE-452/PRC	SFQC	SA	3
331-18E-3044	Construct Sloping-Vee Antenna	SFQC	SA	3
331-18E-3047	Construct Expedient RC-292 Antenna	SFQC	AN	3-4
331-18E-3048	Employ Radio Set AN/PRC-137F	SFQC	SA	3
331-18E-3052	Employ Radio Set AN/PSC-5D	SFQC	QT	3
331-18E-3056	Employ Radio Set AN/PRC-148	ANCOC	QT	3
331-18E-3070	Employ Iridium Emergency Communications	SFQC	SA	3
331-18E-3073	Construct a Slant-Wire Antenna	SFQC	SA	3
Subject Area 4. Communications Planning				
113-611-6002	Plan Frequency Modulated Voice and Data Communications Net	SFQC	AN	3-4
331-18E-3081	Plan for the Employment of Tactical Radio Systems	SFQC	SA	3
331-18E-3083	Install Mission Planning Kit	SFQC	AN	3-4
Skill Level 4				
Subject Area 2. Communication Procedures				
331-18E-4003	Establish a Special Operations Task Force Signal Center	ANCOC	QT	4
Subject Area 4. Communications Planning				
331-18E-4004	Coordinate Signal Activities With Other Units	ANCOC	SA	4
331-18E-4005	Supervise Signal Augmentation	ANCOC	QT	4

STP 31-18E34-SM-TG

Chapter 3

MOS/Skill Level Tasks

Skill Level 3

Subject Area 1: Computer Applications

Install a Tactical Local Area Network (LAN)
113-580-1035

Conditions: As a 31U10 in a tactical operations center, given a network-capable personal computer (PC) compatible or equivalent with manuals, appropriate software with user's manual, cables connectors, hubs, signal operating instructions (SOI), and site map; receive orders to set up the tactical local area network (LAN).

Standards: The standards are met when the tactical LAN is installed and when each station has sent and received a test message.

Performance Steps

1. Review a site map.
    a. Identify network type.
    b. Identify assigned Internet protocol (IP) addressing.
    c. Identify required resources.

2. Inventory resources.

3. Connect hardware/peripherals to the network.
    a. Place stipulated equipment in accordance with (IAW) the site map.
    b. Connect all hardware/peripherals IAW the site map.

4. Initialize PC systems.
    a. Power up hardware.
    b. Verify functionality of PC systems.
    c. Ensure that appropriate operating system (OS) and network software are installed.

5. Configure network software.
    a. Assign IP address to each system as appropriate.
    b. Set up user profiles IAW the unit's SOI/standing operating procedure (SOP).
    c. Test connectivity (ping).

6. Perform an operational check.
    a. Send a test message to all addressees.
    b. Receive an acknowledgement from all addressees.

Evaluation Preparation: Ensure adequate resources are available and a site map and the unit's SOI are provided.

Performance Measures   <u>GO</u>   <u>NO-GO</u>

1. Reviewed the site map.   ___   ___
    a. Identified network type.
    b. Identified assigned IP addressing.
    c. Identified required resources.

STP 31-18E34-SM-TG

Performance Measures   **GO**   **NO-GO**

  2. Inventoried resources.   ——   ——

  3. Connected hardware/peripherals to the network.   ——   ——
      a. Placed stipulated equipment IAW the site map.
      b. Connected all hardware/peripherals IAW the site map.

  4. Initialized PC systems.   ——   ——
      a. Powered up hardware.
      b. Verified functionality of PC systems.
      c. Ensured that appropriate OS and network software were installed.

  5. Configured network software.   ——   ——
      a. Assigned IP address to each system as appropriate.
      b. Set up user profiles IAW the unit's SOI/SOP.
      c. Tested connectivity (ping).

  6. Performed an operational check.   ——   ——
      a. Sent a test message to all addressees.
      b. Received an acknowledgement from all addressees.

Evaluation Guidance: Score the Soldier GO if all steps are passed. Score the Soldier NO-GO if any step is failed. If the Soldier fails any step, show what was done wrong and how to do it correctly. Have the Soldier practice until he can correctly perform the task.

References
    Required   **Related**
    UNIT SOI

STP 31-18E34-SM-TG

Operate a Personal Computer
113-581-1001

Conditions: Given a personal computer (PC) loaded with operating system (OS), application software, and a user's manual.

Standards: The standard is met when the system is set up, operates correctly, and the user can print a test message.

Performance Steps

1. Set up PC.
    a. Use user's manual to inventory peripheral devices (that is, monitor, keyboard, and printer) and connect cables.
    b. Connect cables from peripherals to system unit.
    c. Connect power cables to power source.

2. Check machine to ensure it is operational.
    a. Turn on machine and all peripheral devices.
    b. Wait for desktop to appear on the screen.
    c. Open an application.

3. Print a test page to ensure PC works properly.
    a. Go to START on desktop.
    b. Select SETTINGS, PRINTERS.
    c. Double-click PRINTERS icon.
    d. Select PRINTER, PROPERTIES.
    e. Click on PRINT TEST PAGE button.

Performance Measures   <u>GO</u>   <u>NO-GO</u>

1. Make PC operational.
    a. Connect cables from peripherals to system unit.
    b. Check machine to ensure it is operational.
        (1) Turn on machine and all peripheral devices.
        (2) Start an application.
        (3) Ensure system operates correctly.

2. Print a test page to ensure PC works properly.

STP 31-18E34-SM-TG

Transmit a Digital Image
331-18E-3075

Conditions: Given a digital camera, laptop computer, digital communications equipment, antenna, digital imaging and compression software, interface cables and connectors, and a receiving station, in a field or garrison environment.

Standards: Transmit the digital image to the distant station without loss of intelligible quality.

Performance Steps

1. Capture a digital image.
    a. Set up the camera.
    b. Insert the battery.
    c. Insert compact flash memory card.
    d. Check battery power.
    e. Set date/time.
    f. Select image quality/size.
    g. Compose the picture.
    h. Take the picture.
    i. Check the image.

2. Transfer image to computer.

3. Compress the image.

4. Transmit the image.

Evaluation Preparation: Setup: Provide the Soldier with a digital camera, laptop computer, digital communications equipment, antenna, digital imaging and compression software, interface cables and connectors, and a receiving station.

Performance Measures	GO	NO-GO
1. Captured a digital image.	——	——
2. Transferred image to computer.	——	——
3. Compressed the image.	——	——
4. Transmitted the image.	——	——

Evaluation Guidance: Score the Soldier GO if all performance measures are passed. Score the Soldier NO-GO if any performance measure is failed. If the Soldier fails any performance measure, show what was done wrong and how to do it correctly.

Subject Area 2: Communication Procedures

Communicate in a Radio Net
113-571-1003

Conditions: Given a radio set; ACP 124(D), *Communication Instructions Radiotelegraph Procedure*; ACP 125(F), *Communication Instructions Radiotelephone Procedure*; ACP 126(C), *Communication Instructions Teletypewriter (Teleprinter) Procedures*; FM 24-1, *Signal Support in the Airland Battle*; and unit signal operating instructions (SOI), the student will be aware of contemporary operational environment factors such as extreme temperature differences, unstable political state, and hostile locals, which are the essence to maintain vital communications.

Note: Supervision and assistance are available.

Standards: Establishes, enters, and leaves the radio net in accordance with (IAW) the appropriate references, while observing the contemporary operational environment.

Performance Steps
(Refer to ACP 124(D), ACP 125(F), and ACP 126(C) for performance steps 1 through 3, except as noted.)

1. Establish a radio net.
   a. Extract appropriate call signs, suffixes, and frequency from the unit SOI.
   b. Prepare and operate the appropriate radio set.
   c. Identify the net structure, determine the answering sequence, and make the appropriate response to the individual stations.

2. Enter a radio net. (Refer to ACP 124(D), ACP 125(F), ACP 126(C), and FM 24-1.)
   a. Use abbreviated call signs except when directed by the net control station (NCS) to use full call signs when confusion may result or when entering a net you do not normally operate in.
   b. Authenticate when challenged by NCS.
   c. If you fail to answer a multiple or collective call sign in sequence, wait until all other stations have answered, then answer.
   d. If you are unable to communicate with NCS due to faulty equipment, wrong codes, unsuitable location, and so on, you must render a report to NCS as soon as possible by means other than a radio.

3. Stations leave and close the net.
   a. Request permission from NCS to leave the net.
   b. Inform NCS of the reason for leaving the net.
   c. Authenticate upon direction of NCS prior to leaving the net.

Performance Measures   <u>GO</u>   <u>NO-GO</u>
(Refer to ACP 124(D), ACP 125(F), and 126(C) for performance measures 1 through 3, except as noted.)

1. Established a radio net.   ___   ___
   a. Extract appropriate call signs, suffixes, and frequency from the unit SOI.
   b. Prepare and operate the appropriate radio set.
   c. Identify the net structure, determine the answering sequence, and make the appropriate response to the individual stations. (Refer to Figure 1.)

STP 31-18E34-SM-TG

Performance Measures   GO   NO-GO

LINK STATUS	FEC	FAULT LED	BER LED	SYNC LED	TESTED LED	FEC LED
BER < 10-6	On	Green	Green	Green	Off	Yellow
BER > 10-6	On	Red	Red	Red	Off	Yellow

Figure 1. Net structure

2. Entered a radio net. (Refer to ACP 124(D), ACP 125(F), ACP 126(C), and FM 24-1.) ____ ____
    a. Use abbreviated call signs except when directed by NCS to use full call signs when confusion may result or when entering a net you do not normally operate in.
    b. Authenticate when challenged by NCS.
    c. If you fail to answer a multiple or collective call sign in sequence, wait until all other stations have answered, then answer.

3. Station left and closed the net. ____ ____
    a. Request permission from NCS to leave the net.
    b. Inform NCS of the reason for leaving the net.
    c. Authenticate upon direction of NCS prior to leaving the net.

Evaluation Guidance: Score the Soldier a GO if all performance measures are passed. Score the Soldier a NO-GO if any performance measure is failed. If the Soldier fails any performance measure, show what was done wrong and how to do it correctly. Have the Soldier perform the performance measures until they are done correctly.

References
    Required   **Related**
    FM 24-1

STP 31-18E34-SM-TG

Operate in Radio Nets
113-571-1004

Conditions: Given a requirement and the following: radio set, unit signal operating instructions (SOI), automated network control device (ANCD), applicable technical manual (TM) for radio set used, applicable Army regulations (ARs), and applicable allied communications publications (ACPs). Supervision and assistance will be available.

Standards: Task standard has been met when you have properly entered the selected radio net and authenticated upon request of the net control station (NCS). You have transmitted and received traffic as directed by the NCS, performed duties as NCS, and have left and/or closed the net in accordance with applicable operating procedures and ACP for the radio net in which you are operating according to Performance Measures 1 through 5.

Performance Measures	GO	NO-GO
1. Determined operational net to be entered. (Refer to SOI or ANCD.)	——	——
2. Requested permission to enter net. (Refer to ACP 125(F), *Communication Instructions Radiotelephone Procedure*; ACP 124(D), *Communication Instructions Radiotelegraph Procedure*; and ACP 126(C), *Communication Instructions Teletypewriter (Teleprinter) Procedure*.	——	——
3. Passed traffic as directed by the NCS. (Refer to ACP 125(F), ACP 124(D), and ACP-126(C).)	——	——
4. Requested permission from the NCS to leave the net. (Refer to ACP 125(E); ACP 124(D); ACP 126(C); ACP 131(F), *Communications Instructions Operating Signals*; and station leaving net and closing a net.)	——	——
5. Performed functions of an NCS. (Refer to ACP 125(F), ACP 124(D) [Precedence Prosigns].)	——	——

Evaluation Guidance: Score the Soldier GO if all performance measures are passed (P). Score the Soldier NO-GO if any performance measure is failed (F). If the Soldier fails any performance measure, show what was done wrong and how to do it correctly.

References

Required	Related
ACP 124(D)	
ACP 125(F)	
ACP 126(C)	
ACP 131(F)	

STP 31-18E34-SM-TG

Check Signal Security Procedures
113-573-0001

Conditions: Given a requirement to check signal security (SIGSEC) procedures of an established signal node with organic cryptosystems; Army Regulation (AR) 380-40, *Policy for Safeguarding and Controlling Communications Security (COMSEC) Material*; and the unit operation order (OPORD)/operation plan (OPLAN).

Note: Different types of signal operational requirements will be in effect for this task.

Standards: Checked emission, physical, crypto, transmission, and electronics requirements of security, and took corrective action for any discrepancy noted.

Performance Steps

1. Review the mission OPORD/OPLAN and AR 380-40 to determine specific SIGSEC policies before inspecting the signal node.

2. Check emission security.

3. Check physical security including:
   a. Signal node area of operation.
   b. Area where a cryptosystem is employed.
   c. Emergency evacuation and destruction plans.
   d. Handling of classified material waste.
   e. Control of access and crypto safeguards.

4. Identify physical insecurities.

5. Identify personnel insecurities.

6. Check cryptographic security including:
   a. Proper use of cryptosystems.
   b. Encryption of all classified information.
   c. Competent operation of cryptosystems.

7. Identify crypto insecurities.

8. Check transmission security (TRANSEC) including:
   a. Radio communications.
   b. Conventional telephone communications.

9. Implement appropriate corrective action for any discrepancy noted.

Evaluation Preparation: Setup: Different types of signal operational requirements will be in effect for this task. Brief Soldier: You are required to check SIGSEC at the signal area node and make the necessary corrections.

Performance Measures	GO	NO-GO
1. Reviewed the mission OPORD/OPLAN and AR 380-40 to determine specific SIGSEC policies before inspecting the signal node.	——	——
2. Checked emission security.	——	——

3. Checked physical security including:
  a. Signal node area of operation.
  b. Area where a cryptosystem is employed.
  c. Emergency evacuation and destruction plans.
	——	——

Performance Measures <u>GO</u>   <u>NO-GO</u>
      d. Handling of classified material waste.
      e. Control of access and crypto safeguards.
      f. Identify physical insecurities.

4. Identified physical insecurities.   ____   ____

5. Identified personnel insecurities.   ____   ____

6. Checked cryptographic security including: ____   ____
      a. Proper use of cryptosystems.
      b. Encryption of all classified information.
      c. Competent operation of cryptosystems.

7. Identified crypto insecurities.   ____   ____

8. Checked TRANSEC including: ____   ____
      a. Radio communications.
      b. Conventional telephone communications.

9. Implemented appropriate corrective action for any discrepancy noted.   ____   ____

Evaluation Guidance: Score the Soldier GO if all performance measures are passed (P). Score the Soldier NO-GO if any performance measure is failed (F). If the Soldier fails any performance measure, show what was done wrong and how to do it correctly.

References
    Required   **Related**
    AR 380-40   AR 380-5
                                   AR 530-1
                                   FM 3-19.30

Conduct Communications Security Inspections
113-573-1004

Conditions: Given Allied Communication Publication (ACP) 124(D), *Communication Instructions Radiotelegraph Procedure*; ACP 125(F), *Communication Instructions Radiotelephone Procedure*; ACP 126(C), *Communication Instructions Teletypewriter (Teleprinter) Procedure*; Army Regulation (AR) 380-5, *Department of the Army Information Security Program*; Technical Bulletin (TB) 380-41, *Security Procedures for Safeguarding, Accounting, and Supply Control of Communications Security Material*; unit signal operating instructions (SOI), and inspection checklist.

Standards: The standards are met when an operations security inspection (appropriate per unit/mission) is completed and results are recorded and submitted to proper authorities.

Evaluation Preparation: Setup: Different types of signal requirements will be in effect for this task. Brief Soldier: You are required to complete operations security inspection and make the necessary corrections.

Performance Measures	GO	NO-GO

1. Checked physical security. (Refer to TB 380-41.)
   a. Checked defense of restricted areas (fences, guards, dogs, and so forth).
   b. Checked access roster.
   c. Checked doors, locks, and windows.
   d. Checked Standard Form (SF) 702 (Security Container Check Sheet), SF 700 (Security Container Information), Department of the Army (DA) Form 2653-R (COMSEC Account—Daily Shift Inventory), and SF 153 (COMSEC Material Report).
   e. Checked emergency evacuation.
   f. Checked destruction records.
   g. Checked the reporting of violation suspected or actual.
	——	——

2. Checked transmission security. (Refer to ACP 124(D), ACP 125(F), ACP 126(C), and unit SOI.)
   a. Ensured 20-second transmission time is being enforced.
   b. Ensured messages are preplanned before transmitting.
   c. Ensured equipment is operated in the secure mode.
	——	——

3. Checked cryptographic security utilizing the crypto facility checklist in TB 380-41.	——	——
4. Submitted completed results of the inspection to proper authorities.	——	——

Evaluation Guidance: Score the Soldier GO if all performance measures are passed (P). Score the Soldier NO-GO if any performance measure is failed (F). If the Soldier fails any performance measure, show what was done wrong and how to do it correctly.

Conduct Shift-to-Shift Inventory of Communications Security Material
113-573-2029

Conditions: Given communications security (COMSEC) material for inventory and Department of the Army (DA) Form 2653-R (COMSEC Account—Daily Shift Inventory).

Standards: The standards are met when COMSEC material is accounted for and DA Form 2653-R is complete.

Performance Steps

1. The COMSEC custodian is responsible for the accuracy of the inventory performed under his supervision and MUST SIGN the inventory report on the COMSEC custodian line. The selected witness will also sign this inventory report.
2. The following procedures are required for all COMSEC material within the COMSEC material control system (CMCS), as received at COMSEC logistics support facilities (CLSFs) for further distribution:
   a. Carefully examine packages and containers for signs of tampering. If there are signs of tampering, refer to Technical Bulletin (TB) 380-41, *Security Procedures for Safeguarding, Accounting, and Supply Control of Communications Security Material,* for procedures. If there are no signs of tampering, do not open the package or container. Inventory of unopened packages and containers is done by verifying the short title/edition, quantity, and serial (accounting)/register number on the label.
   b. Larger packages, such as bulk-packaged COMSEC key material, may have been opened for distribution of the smaller unit-packed contents, such as key cards. If the large package is open, inventory its contents. Large packages opened for distribution of smaller units may be resealed and remarked with the new quantity. There is no need to open smaller packages for inventory unless they are open or show signs of tampering. If they are open or show signs of tampering, verify that the contents are complete.
   c. Do not open spare component packages for inventory unless there are signs of tampering. However, if the package is open, inventory all individual classified components.
3. Periodic inventory.
   a. Inventory the large sealed units. Large sealed units may include one or more pallets of material as a total sealed unit. Check all seals for signs of tampering.
   b. Inventory the material, which is stored loose, such as accounting legend code (ALC) 1 and ALC 3 keying material. Do not check equipment and subassemblies for individual parts.
4. When COMSEC material is transferred, check both the equipment and individual components for completeness before they are packed for shipment. If the equipment or components have not been opened, you are not required to open and check these items.
5. Daily and shift-to-shift inventory of key material.
   a. All CRYPTO-marked key will be inventoried by the holders to assure continuing protection and control.
   b. Secure any classified key not required for daily use in vaults or containers. Access to the combination will be limited to the COMSEC custodian and alternate COMSEC custodian(s).
      (1) A daily inventory will be conducted in all accounts holding classified key material, regardless of quantity held. DA Form 2653-R) will be used for this inventory. The form will be locally reproduced on 8 x 11-inch paper. Key stored in security containers that have not been opened since the previous inventory need not be opened for the purpose of conducting a daily inventory.
      (2) DA Form 2653-R is maintained as a matter of record IAW TB 380-41. Page checks of unsealed key are not required for daily inventories.

STP 31-18E34-SM-TG

Performance Steps

    c. Shift-to-shift inventory. Facilities engaged in continuous operations will conduct shift-to-shift inventories. During the inventory of partially used keying material, personnel will ensure the combination of on-hand keying material and local records of destruction account for the entire item.
    d. Daily inventory record. Record daily and shift-to-shift inventories on DA Form 2653-R. Data processing lists may be used for shift-to-shift inventories provided they contain the same information as the DA Form 2653-R.
    e. The instructions for completing DA Form 2653-R are as follows:
        (1) **SHORT TITLE** - Enter the short title and edition number for each item being inventoried.
        (2) **QTY** - Enter the required inventory quantity.
        (3) **REG NO.** - Enter the keying material register number or serial number (SN).
        (4) **SHIFT** - Use the appropriate shift row (shift 1, shift 2, or shift 3) to record your inventory.
        (5) **DAY OF THE MONTH** - Use the correct day of the month column to record your inventory.
        (6) **INVENTORY RECORD** - Select the box that intersects your shift and the day of the month. If the inventory count agrees with the required quantity, place a check mark (/) in the inventory record.
        (7) **INITIALS** - If the inventory of all listed key material items is correct, place your initials in the box which corresponds to the correct shift (1, 2, or 3) and the day of the month.
        (8) **MONTH** - Enter the month for which the inventory is being performed.
        (9) **PAGE NUMBER** - Number each sheet consecutively.
        (10) **NUMBER OF PAGES** - Enter the total number of pages included for the entire month.
        (11) **DELETIONS** - Will be noted as shown in Figure A-1 in Appendix A.
        (12) **ADDITIONS** - Will be noted as shown in Figure A-1 in Appendix A.

Evaluation Preparation: Setup: NA. Brief Soldier: Tell the Soldier to conduct a shift-to-shift inventory of COMSEC material and complete DA Form 2653-R.

Performance Measures	GO	NO-GO

1. Inventoried COMSEC material and completed DA Form 2653-R.
   a. Entered the short title for each item inventoried.
   b. Entered the quantity.
   c. Entered the keying material register number or serial number.
   d. Used the appropriate shift row.
   e. Used the correct day of the month.
   f. Selected the box, which intersects your shift and the day of the month.
   g. Entered the month.
   h. Numbered each sheet consecutively.
   i. Entered the total number of pages.
	——	——

2. Ensured initials are placed in the correct block.	——	——
3. Assigned the appropriate classification to the DA Form 2653-A.	——	——

Evaluation Guidance: Score the Soldier GO if all performance measures are passed (P). Score the Soldier NO-GO if any performance measure is failed (F). If the Soldier fails any performance measure, show what was done wrong and how to do it correctly.

References

Required	Related
	AR 380-5

STP 31-18E34-SM-TG

Recognize Electronic Attack and Implement Electronic Protection
113-573-6001

Conditions: Given a radio set; applicable operator's technical manual (TM); Field Manual (FM) 24-1, *Signal Support in the Airland Battle*; and unit signal operating instructions (SOI) extract.

Standards: Determine that electronic warfare (EW) is directed at your station and electronic counter-countermeasures (ECCM) are employed for continued operation.

Performance Steps

1. Introduction. A close relationship exists between ECCM and communications security (COMSEC). Both defensive arts are based on the same principle. An enemy who does not have access to our essential elements of friendly information (EEFI) is a much less effective foe. The major goal of COMSEC is to ensure that friendly use of the electromagnetic spectrum for communications is by the enemy. The major goal of practicing sound ECCM techniques is to ensure the continued use of the electromagnetic spectrum. ECCM techniques are designed to ensure commanders some degree of confidence in the continued use of these techniques. Our objective must be to ensure that all communications equipment can be employed effectively by tactical commanders in spite of the enemy's concerted efforts to degrade such communications to the enemy's tactical advantage. The modification and the development of equipment to make our communications less susceptible to enemy exploitation are expensive processes. Equipment is being developed and fielded which will provide an answer to some of ECCM problems. Commanders, staff, planners, and operators remain responsible for security and continued operation of all communications equipment.
    a. Operators of communications equipment must be taught what jamming and deception can do to communications. They must be made aware that incorrect operating procedures can jeopardize the unit's mission and ultimately increase unit casualties. Preventive and remedial ECCM techniques must be employed instinctively. Maintenance personnel must be made aware that unauthorized or improperly applied modifications may cause equipment to develop peculiar characteristics which can be readily identified by the enemy.
    b. ECCM should be preventive in nature. ECCM should be planned and applied to force the enemy to commit more jamming, interception and deception resources to a target than it is worth, or is available. ECCM techniques must also be applied to force the enemy to doubt the effectiveness of the enemy's jamming and deception efforts.
    c. Before we can begin to prevent electronic countermeasures (ECM), we must first be certain of what we are trying to prevent.
        (1) Jamming is the deliberate radiation, reradiation, or reflection of electromagnetic energy with the object of impairing the use of electronic devices, equipment, or systems. The enemy conducts jamming operations against us to prevent us from effectively employing our radios, radars, navigational aids (NAVAIDS), satellites, and electro-optics. Obvious jamming is normally very simple to detect. The more commonly used jamming signals of this type are described below. Do not try to memorize them; just be aware that these and others exist. When experiencing a jamming incident, it is much more important to recognize it and take action to overcome it than to identify it formally.
            (a) *Random noise.* It is random in amplitude and frequency. It is similar to normal background noise and can be used to degrade all types of signals.
            (b) *Stepped tones.* These are tones transmitted in increasing and decreasing pitch. They resemble the sound of bagpipes.
            (c) *Spark.* The spark is easily produced and is one of the most effective forms of jamming. Bursts are of short duration and high intensity. Sparks are repeated at a rapid rate and are effective in disrupting all types of communications.
            (d) *Gulls.* The gull signal is generated by a quick rise and a slow fall of a variable radio frequency and is similar to the cry of a sea gull.

STP 31-18E34-SM-TG

Performance Steps

        (e) *Random pulse.* In this type of interference, pulses of varying amplitude, duration, and rate are generated and transmitted. Random pulses are used to disrupt teletypewriter, radar, and all types of data transmission systems.
        (f) *Wobbler.* The wobbler is a single frequency which is modulated by a low and slowly varying tone. The result is a howling sound which causes a nuisance on voice radio communications.
        (g) *Recorded sounds.* Any audible sound, especially of a variable nature, can be used to distract radio operators and disrupt communications. Examples of sounds include: music, screams, applause whistles, machinery noise, and laughter.
        (h) *Preamble jamming.* This type of jamming occurs when the synchronization tone of speech security equipment is broadcast over the operating frequency of secure radio sets. Preamble jamming results in radios being locked in the receive mode. It is especially effective when employed against radio nets using speech security devices.
        (i) *Subtle jamming.* This type of jamming is not obvious at all. With subtle jamming, no sound is heard from our receivers. They cannot receive incoming friendly signals, but everybody appears normal to the radio operator.
    (2) *Meaconing.* This is a system of receiving radio beacon signals from NAVAIDS and rebroadcasting them on the same frequency to confuse navigation. The enemy conducts meaconing operations against us to prevent our ships and aircraft from arriving at their intended targets or destinations.
    (3) *Intrusion.* Intentional insertion of electromagnetic energy into transmission paths with the objective of deceiving equipment operators or causing confusion. The enemy conducts intrusion operations against us by inserting false information into our receiver paths. This false information may consist of voice instructions, ghost targets, coordinates for fire missions, or even rebroadcasting of prerecorded data transmissions.
    (4) *Interference.* Interference is any electrical disturbance which causes undesirable responses in electronic equipment. As a meaconing, interference, jamming, and intrusion (MIJI) term, interference refers to the unintentional disruption of the use of radios, radars, NAVAIDS, satellites, and electro-optics. This interference may be of friendly, enemy, or atmospheric origin. For example, a civilian radio broadcast interrupting military communications is interference.

2. Communications Protective Measures.
    a. Considerations. Properly applied ECCM techniques will deny valuable intelligence sources to the enemy and eliminate much of the threat that he poses to our combat operations. The following discussion describes practical ways to protect communications systems.
    b. The siting of the transmitting antenna is critical in the ECCM process. Before making a decision about a proposed site for either a single-channel or multichannel antenna, there are two basic questions to answer:
        (1) Are communications possible from the proposed site?
        (2) Are there enough natural obstacles between the site and the enemy to mask transmission?
    c. The final decision on site selection will often be a tradeoff between the answers to these two questions. The communications mission must have first priority in determining the actual antenna sites. There are additional actions that must be taken to limit the enemy's chances of interception and location successes. Transmitters and antennas should be located away from the headquarters. The two locations should be separated by more than 1 kilometer (0.62 mile). Erroneous radio frequency direction (RFD) data used in conjunction with observation data may favor the targeting of a decoy site instead of the actual transmitter site. This ploy depends upon good camouflage at the actual site. Transmitters grouped in one area indicate the relative value of the headquarters. Directional antennas reduce radiation exposure to enemy receivers and enhance the intended signal. (For instruction on directional antennas, refer to TC 24-21, *Tactical Multichannel Radio Communications Techniques*.)
    d. Use the lowest possible transmitter power output. Less power means less radiated power reaches the enemy and thus increases his difficulty in applying ECM.

STP 31-18E34-SM-TG

Performance Steps

  e. Use only approved code systems. Never use unauthorized (homemade) codes. Use of non-National Security Agency (NSA) generated codes can provide a false COMSEC sense of security that can be exploited by enemy radio intercept operators. Only when absolutely necessary should traffic be passed in the clear.
  f. Rather than assuming equipment is defective, assume that it is operational. Operators must not contact other stations for equipment checks simply because no message has been transmitted in a set time frame.

Evaluation Preparation: Setup: A radio set operating in a radio net with interference applied to the system. Brief Soldier. Tell the Soldier to apply proper tactics to the jamming system.

Performance Measures	GO	NO-GO

1. Determined if ECM was being employed.
    a. Checked for accidental or unintentional interference.
    b. Checked for intentional interference.
	——	——

2. Initiated operator's procedures. (Refer to FM 24-1.)
    a. Checked the equipment ground to ensure that the interference was not caused by a buildup of static electricity.
    b. Disconnected the antenna.
    c. Identified the type of sound.
    d. Moved the receiver or reoriented the antenna, if possible, and listened or looked for variations in the strength of the disturbance.
    e. Tuned the receiver above or below the normal frequency. If such detuning caused the intensity of the interfering signal to drop sharply, it can be assumed that the interference was the result of spot jamming.
	——	——

3. Identified jamming signals.	——	——
4. Employed antijamming measures. (Refer to FM 24-1.)	——	——

Note: Antijamming measures are designed to allow radio operators to work effectively through intentional interference. Regardless of the nature of the interfering signal, radio operators WILL NOT reveal in the clear the possibility or success of enemy jamming.

Evaluation Guidance: Score the Soldier GO if all performance measures are passed (P). Score the Soldier NO-GO if any performance measure is failed (F). If the Soldier fails any performance measure, show what was done wrong and how to do it correctly.

References
  Required   **Related**
                                        FM 24-1
                                        TC 24-21

STP 31-18E34-SM-TG

Use an Automated Signal Operation Instruction
113-573-8006

Conditions: Given signal operating instructions (SOI) KTV 1600, a radio, an operational radio net, scratch paper, and a pencil.

Standards: This task has been performed correctly when the Soldier does the following in 10 minutes:
1. Lists the item number of the SOI extract.
2. Lists a radio station call sign.
3. Lists a radio net frequency.
4. Lists an item number identifier.
5. Enters a radio net in which you do not normally operate.
6. Lists a challenge and reply authentication.

Performance Steps

1. List an SOI item number.
   a. Get the SOI/extract for your unit.
   b. Find the item number for the unit you want (Figure 1). Look down the left-hand column to find the unit *(2ND, BDE, 1-80 IN BN)*. Then look immediately to the right of the unit to find the item number *(8C)*.

KTV 1600C	(PROTECTIVE MARKING) INDEX	1
ITEM		**ITEM NO**
1-77 IN BN		7B
1-78 IN BN		7C
2ND BDE		8
1-3 AR BN		8A
1-79 IN BN		8B
1-80 IN BN		8C
3RD BDE		9
1-4 AR BN		9A
1-81 IN BN		9B
1-82 IN BN		9C
DIVARTY		10
1-40 FA BN		10A
1-41 FA BN		10B
1-42 FA BN		10C
1-43 FA BN		10D
DISCOM		11
52ND MED BN		11A
52ND S & T BN		11B
52ND MAINT BN		11C
1-23 CAV SQDN		12
1-23 D/CAV SQDN		12A
1-441 ADA BN		13
INDEX	2 of 5	1

Figure 1. SOI KTV 1600C

   c. Turn to the item number page for the time period you are using (Figure 2). The item number is located at the upper and lower right of the page. The time period is located at the top right of the page. *Example: Time Period 01, Item number 8C.*

3-16   8 February 2010

Performance Steps

```
                    (PROTECTIVE MARKING)        01
                         TIME PERIOD                        8C
   1-80 IN BN EXTRACT
   1-80 IN BN          Z6Z     57.85   CMD   COMMANDER         37
                               49.65   A/L   XO                26
   HHC/1-80 IN         R1L                   S1                59

   A/1-80 IN           Q3V     36.05   CMD   S2                65
   1/A/1-80            I8C     51.40         S3                89
   2/A/1-80            W1J     50.20         S4/SUP SGT        45
   3/A/1-80            C4K     51.20         MTR OFF/SGT       29
   WPNS/A/1-80         S6Y     65.60         C-E O/COMM CH     25
   B/1-80 IN           I3B     56.65   CMD   MED OFF/MEDIC     95
   1/B/1-80            P6P     50.10         FO 4              18
   2/B/1-80            LOB     51.00         FO 5              78
   3/B/1-80            S3I     49.90         FO 6              06
   WPNS/B/1-80         E4V     34.55         FSO/FIST CHIEF    10
   C/1-80 IN           U4P     44.25   CMD   NCS/TOC/CP        08
   1/C/1-80            M8O     50.60         FDC               16
   2/C/1-80            Y6M     51.70         PLT/SEC/TM LDR    44
   3/C/1-80            V3X     49.60         PLT/SEC/TM SGT    98
   WPNS/C/1-80         Z4N     34.85         TM/SQD/SEC 1      63
   CSC/1-80 IN         Q6C     68.15   CMD   TM/SQD/SEC 2      51
   AD/CSC/1-80         N2L     45.95         TM/SQT/SEC 3      77
   AT/CSC/1-80         B8G     37.25         TM/SQT/SEC 4      69
   SCT.CSC/1-80        V1R     47.40         TACP              31
   MORT/CSC/1-80       K9T     33.50         MAINT OFF         58
   GSR/CSC/1-80        C9A     61.05         ENGR OFF          41
   1-80 AJ/ALTN 1              74.20         CSM/SGM/1sg       40
   1-80 AJ/ALTN 2              32.25         OFF ASST          H
   2 BDE               X8A     40.05   CMD   ENL ASST          O
   2 BDE RETRANS       T7N     63.40   RTS   RTO/DRIVER        F
   MEDEVAC (P)         W5X     32.75   MED   SIGN              HITS
                                             C/SIGN            YEAR
   1-80 IN BN EXTRACT                                          8C
```

Figure 2. 1-80 Infantry Battalion Extract

2. List a radio station call sign (Figure 2).
    a. Look down the left-hand column to find the unit. *Example: Weapons platoon of Company C, 1-80 Infantry Battalion (WPNS/C/1-80).*
    b. Then look immediately to the right of the unit to find the call sign *(Z4N).*
    c. Find the suffix that designated the person or subordinate element of the unit by reading down the list of suffixes on the right side of the page and reading the two-digit suffix for that element *(the suffix for PLT LDR is 44).* Add it to the previous call signs you found. You now have the five-character call sign for the element/person *(Z4N44).*

3. List a radio net frequency (Figure 2). Look to the immediate right of your call sign. You now have the opening frequency of the weapons platoon *(34.85 megahertz [MHz]).*

4. List an item number identifier.
    a. With the item number for your unit (paragraph 1) turn to the item number identifiers section of your SOI/extract (see Figure 3).

STP 31-18E34-SM-TG

Performance Steps

KTV 1600C				(PROTECTIVE MARKING) ITEM NUMBER IDENTIFIERS						23
	01	02	03	04	05	06	07	08	09	10
CA	9C	14	11B	10	16G	10B	11A	21	16G	3D
CB	16A	16J	13A	9C	7B	5A	22B	9C	12	4A
CC	16E	22D	16C	8B	9B	6A	21	3F	22E	3
CD	3B	22B	16E	7A	8B	7B	13	16B	11C	9B
CE	10	16	15	16D	18	13	3A	7C	21	12
CF	8C	16A	16F	22E	3	4A	16A	6B	12A	7B
CG	9A	13	9B	16I	22B	3B	6D	6A	19	10A
CH	22B	16G	5	16G	22	6C	10	16A	16C	11B
CI	13	3F	3A	19	16B	8C	5A	16D	18	9
CJ	7B	22A	11C	4A	10C	22B	8	6	4	16G
CK	3E	3A	16G	3A	6B	22E	6	8	6	4
CL	8	6	21	13A	16J	8	16C	3E	4A	16I
CM	19	16D	22C	16A	9C	16E	16E	13A	22C	3F
CN	21	11A	20	8A	20	3	16I	22	11A	22A

Figure 3. Item Number Identifiers

    b. Find the column for the period you are in *(time period 01)*. The time periods 1-10 are the column heading across the top of the sheet.
    c. Read down this time period column unit you find the item number for your unit.
    Example: The Item Number for 1-80/In BN is 8C.
    d. Read the two-letter code in the left column opposite your unit item number *(CF)*.

5. Enter a radio net in which you do not normally operate.
    a. Turn to the *Quick Ref Maj Subor Elms & CBT BNS* items of your SOI/extract. There are two sets, one for call signs (Figure 4) and one for frequencies (Figure 5).

KTV 1600C QUICK REF MAJOR SUBOR ELMS & CBT BNS		(PROTECTIVE MARKING) CALL SIGNS				3
	01	02	03	04	05	
52D DIV	K6P	D4J	N3D	Y2E	L5D	
1 BDE	V4Y	M8R	Z4S	H1S	X2E	
2 BDE	X8A	Z6N	K2J	K7H	V9A	
3 BDE	V8K	B1P	V8Q	K8W	R4I	
DIVARTY	C9L	X0V	E4H	S2B	A0H	
DISCOM	T6I	Q0Q	U4F	V6R	R1V	
1-2 AR BN	R4S	B8W	A7G	H4I	D9L	
1-3 AR BN	X3W	Z9S	L6P	H2M	S4K	
1-4 AR BN	X4R	R7X	T1B	I9F	C1Y	
1-40 FA BN	D2U	E1F	W4W	A8Q	A5P	
1-41 FA BN	S0D	Z6T	W0M	Z6X	H1X	
1-42 FA BN	T0X	S5D	Q2Y	P8A	S2R	
1-43 FA BN	G2G	L3G	J7I	Q0L	Q9Q	
1-77 IN BN	A1H	N8L	V8R	W4T	X2S	
1-78 IN BN	Q0F	M7I	Z2C	A3K	G1M	
1-79 IN BN	K7O	Z5K	D5N	G7C	N8F	
1-80 IN BN	Z6Z	H6H	M4V	K2U	M9N	
1-81 IN BN	J5N	X9E	R5U	L1Z	Y2W	
1-82 IN BN	V5E	C6U	R2T	D8N	G7U	
1-23 CAV SQDN	N0C	C8Y	N0L	E7O	F5O	
1-441 ADA BN	Y7B	M2B	U7E	J2G	V1C	
52D ENBN	D6J	S1Z	D6X	D4Y	B9T	
52D SIG BN	P4V	G4A	K7Z	B9P	R8B	
52D CAB	J5T	L6C	R5O	M3V	D4Z	
312 CEWI BN	J3Q	P8O	M4A	U2J	G2G	
MEDEVAC	W5X	L5U	X9S	Z7X		
52D MP CO	S8M	Y4M	T8K	U4D	R8J	
QUICK REF MAJOR SUBOR ELMS & CBT BNS						3

Figure 4. Quick Reference: Major Subordinate Elements and Combat Battalions—Call Signs

STP 31-18E34-SM-TG

Performance Steps

KTV 1600C QUICK REF MAJOR SUBOR ELMS	(PROTECTIVE MARKING) FREQUENCIES ELMS & CBT BNS					3
	01	02	03	04	05	
DIV CMD	38.05	63.75	68.10	65.45	59.15	
1BDE CMD	51.65	68.30	46.80	62.70	64.65	
2 BDE CMD	40.05	50.50	39.55	69.50	51.50	
3 BDE CMD	55.05	67.50	44.35	63.20	69.85	
DIVARTY CF 1	54.50	69.20	60.90	42.90	67.00	
DISCOM CMD	49.95	34.05	64.45	58.10	56.00	
1-2 AR CMD	34.65	37.30	66.95	44.90	64.45	
1-3 AR CMD	46.25	31.70	66.70	52.90	55.65	
1-4 AR CMD	66.85	40.30	60.15	40.90	69.55	
1-40 FA CF	61.90	36.60	49.85	43.60	38.45	
1-41 FA CF	56.05	46.50	61.70	39.35	58.35	
1-42 FA CF	34.30	65.30	62.75	60.80	34.45	
1-43 FA CF	58.20	49.15	64.00	48.90	51.60	
1-77 IN CMD	44.20	46.00	41.65	32.35	50.05	
1-78 IN CMD	32.65	31.95	54.50	62.45	62.75	
1-79 IN CMD	47.75	53.10	52.35	39.45	53.15	
1-80 IN CMD	57.85	55.35	30.55	64.60	66.35	
1-81 IN CMD	48.55	35.15	39.90	67.30	35.85	
1-82 IN CMD	38.40	48.40	62.95	33.35	47.85	
1-23 CAV CMD	43.10	47.90	42.40	37.90	62.15	
1-441 ADA CMD	67.65	38.65	58.25	46.35	37.45	
52D EN CMD	51.60	47.55	47.40	31.60	63.45	
52D SIG CMD	32.80	46.70	51.10	58.20	34.35	
52D CAB CMD	54.70	54.95	35.75	53.35	34.05	
312 CEWI BN CMD	30.65	55.20	38.90	63.40	69.95	
MEDEVAC P	32.75	32.75	32.75	32.75	32.75	
MEDEVAC A	53.55	55.40	55.55	40.40	46.85	
QUICK REF MAJOR SUBOR ELMS ELMS & CBT BNS						3

Figure 5. Quick Reference: Major Subordinate Elements and Combat Battalions—Frequencies

    b. Look down the left-hand column of the call signs set to find the unit *(1-3 AR BN)*. Then look immediately to the right of the unit to find the call sign under the correct time period column. *Example: 1-3 AR BN under Time Period 01, the call sign is X3W (Figure 4).*

    c. Turn to the frequencies set and repeat the procedures to find the frequency of the battalion command net *(1-3 AR BN, Time Period 01, 46.25MHz) (Figure 5).*

    d. Set your radio to the frequency for the net control station (NCS).

    e. Call the NCS and request permission to enter the net. **Example:** *Call:* X-RAY THREE WHISKEY ZERO EIGHT, THIS IS ZULU FOUR NOVEMBER FOUR, REFER TO CHARLIE. (CF is the Number Identifier for your unit). *I HAVE TRAFFIC FOR X-RAY THREE WHISKEY EIGHT NINER. REQUEST PERMISSION TO ENTER YOUR NET, OVER.*

    f. Give the correct reply when the NCS challenges. **Example:** (Challenge by the NCS.) *ZULU FOUR NOVEMBER FOUR, THIS IS X-RAY WHISKEY ZERO EIGHT. AUTHENTICATE CHARLIE HOTEL, OVER.* (Reply by you or the caller). *X-RAY THREE WHISKEY ZERO EIGHT, THIS IS ZULU FOUR NOVEMBER FOUR. I AUTHENTICATE LIMA, OVER.*

Note: The station being called will make the first challenge. Both stations must find the correct reply so that the station being called can authenticate the reply by the calling station. If the called station does not respond to the challenge within a reasonable time, the station calling will require another authentication using different challenge. Either station can challenge the other if there is a reason to believe that the other station is not a friendly station.

    g. When the NCS grants permission to enter the net, find the call sign for the unit you want.

    h. Call the unit you want and send your message.

    i. After you finish sending your message, call the NCS and ask to leave the net. You should be required to authenticate.

STP 31-18E34-SM-TG

Performance Steps

6. List a challenge and reply authentication.
   a. Get the KTC 1400 section of your SOI.
   b. Turn to the set (page) for the time period you are using *(01)* (Figure 6).

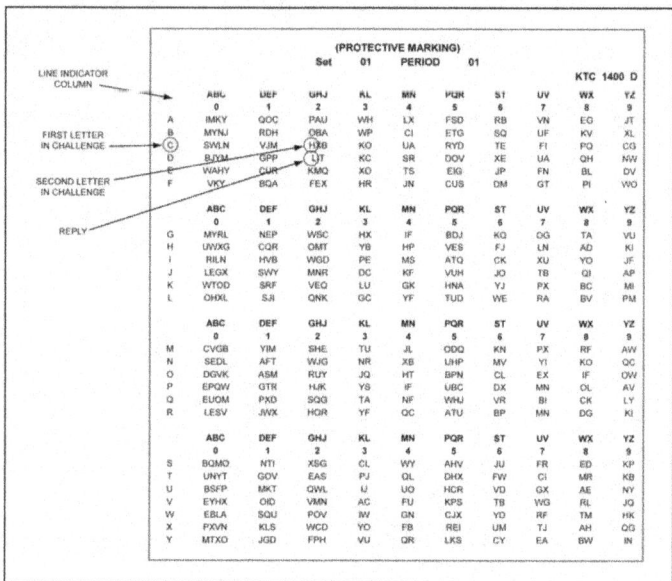

Figure 6. Set page

  c. The challenge is any two letters (except Z), selected at random. **Example:** *"C H"*.
  d. Find the first letter of the challenge *("C")* in the line indicator column on the left of the page.
  e. Read to the right on that line to find the second letter *("H")*.
  f. Read the letter directly under the second letter *("L")*. This is the correct reply to the challenge. If the first letter is *"Y"*, go to the top of the sheet in the same column to find the reply.
  Example: *for challenge "YR", the reply is "X"*.

Evaluation Preparation: Setup: Provide the Soldier with an SOI KTV 1600D, KTC 1400 a radio, an operational radio net, scratch paper, and a pencil. Brief Soldier: Tell the Soldier to perform each step correctly and to complete it within 10 minutes.

Performance Measures   GO   NO-GO

  1. Listed the Item Number of the SOI extract.   ____   ____

  2. Listed a radio net frequency.   ____   ____

  3. Listed a radio station call sign.   ____   ____

  4. Listed an item number identifier.   ____   ____

  5. Entered a radio net in which you do not normally operate.   ____   ____

STP 31-18E34-SM-TG

Performance Measures   <u>GO</u> <u>NO-GO</u>

  6. Listed a challenge and reply authentication. —— ——

Evaluation Guidance: Score the Soldier GO if all performance measures are passed (P). Score the Soldier NO-GO if any performance measure is failed (F). If the Soldier fails any performance measure, show what was done wrong and how to do it correctly.

References
    Required   **Related**
    UNIT SOI

STP 31-18E34-SM-TG

Destroy Classified Material
113-573-9012

Conditions: Given Army Regulation (AR) 380-5, *Department of the Army Information Security Program*; Department of the Army (DA) Form 3964 (Classified Document Accountability Record) or a certificate of destruction (as appropriate); destruction facility or equipment; witness; appropriate moving storage bags/boxes; transportation (if required); and TB 380-41, *Security Procedures for Safeguarding, Accounting, and Supply Control of Communications Security Material*.

Standards: The standards are met when the documentation is complete and the classified material is destroyed by burning or shredding.

Performance Steps

1. When dealing with classified material, it is important that all guidelines, Army regulations, and installation/unit polices be followed.

Note: In some instances, classified material will include equipment such as hard drives, disk packs, and tape cartridges. Be sure the S-4 office or unit supply is well aware that equipment is to be destroyed and ensure the properly completed supply documents accompany the equipment.

2. Several things must be accomplished when you are tasked to destroy classified documents. You must know the classification of the documents to be destroyed. The classification determines the paperwork needed, the security clearances required, method of destruction required, witness, and so on.

Note: If you are the witness, the person in charge of the required destruction will decide what items require your assistance.

   a. Safeguarding—When tasked to destroy information classified under AR 380-5, that information shall be afforded the level of protection against unauthorized disclosure commensurate with the level of classification assigned under the varying conditions that may arise in connection with its use, dissemination, storage, movement or transmission, and destruction. Responsible officials will ensure classified information is adequately protected from compromise. Officials must be continually aware of possible threats from all-source intelligence efforts of potential adversaries (AR 380-5, paragraph 1-402).
   b. Classified information access policy—No person may have access to classified information unless that person has been determined to be trustworthy and unless access is essential to the accomplishment of lawful and authorized government purposes, that is, the person has the appropriate security clearance and a need-to-know. No one has a right to have access to classified information solely by virtue of rank or position.

3. Destruction of classified documents and material--These procedures shall incorporate means of verifying the destruction of classified information and material (AR 380-5, paragraph 9-100).
   a. Classified documents and material shall be destroyed by burning or, with the approval of the cognizant Department of Defense (DOD) component head or designee, by melting, chemical decomposition, pulping, pulverizing, cross-cut shredding, or mutilation sufficient to preclude recognition or reconstruction of the classified information, (TB 380-41). In all cases, burning is the preferred method of destroying classified information (AR 380-5, paragraph 9-101).
   b. Procedures shall be instituted that ensure all classified information intended for destruction actually is destroyed. Destruction records and imposition of a two-person rule, that is, having two cleared persons involved in the entire destruction process, will satisfy this requirement for TOP SECRET information. Imposition of a two-person rule, without destruction records, will satisfy this requirement for SECRET information, as will use of destruction records without imposition of the two-person rule. Only one cleared person needs to be involved in the destruction process for CONFIDENTIAL information (AR 380-5, paragraph 9-102).

Performance Steps

    c. Records of destruction (DA Form 3964) are required for TOP SECRET information. The record shall be dated and signed at the time of destruction by two persons cleared for access to TOP SECRET information. Records of destruction of SECRET and CONFIDENTIAL information are not required except for NATO SECRET and some limited categories of specially controlled SECRET information. When records of destruction are used for SECRET information, only one cleared person has to sign such records. DA Form 3964 will normally be used as the record of destruction (AR 380-5, paragraph 9-103).

    d. Waste materials such as handwritten notes, carbon paper, typewriter ribbons, and working papers that contain classified information must be protected to prevent unauthorized disclosure of the information. Classified waste shall be destroyed when no longer needed. Destruction records are not required (AR 380-5, paragraph 9-104).

4. Destruction of classified hardware equipment--Approval of destruction methods will be obtained from United States Army Intelligence and Security Command (INSCOM) prior to destruction. Usually there will be an installation security office to advise you or do the destruction for you. Only under EMERGENCY CONDITIONS OCONUS is destruction or other disposition of equipment components by the individual user or user organization authorized.

Note: All waste material generated within a cryptofacility (worksheets, tape, carbon paper, typing ribbons, and so forth) both classified and unclassified, will be disposed of in the same manner as directed by AR 380-5 for the destruction of classified waste.

5. Penalties for violation of security regulations—DOD military and civilian personnel are subject to administrative sanctions if they knowingly, willfully, or negligently disclose to unauthorized persons information properly classified under Executive Order 12356, *National Security Information;* prior orders; any implementing issuances; or AR 380-5. Sanctions include but are not limited to a warning notice, reprimand, termination of classification authority, suspension without pay, forfeiture of pay, removal or discharge, and will be imposed upon any person, regardless of office or level of employment, who is responsible for a violation specified under this paragraph as determined appropriate under applicable law and DOD regulations. Nothing in AR 380-5 prohibits or limits action under the Uniform Code of Military Justice based upon violations of that code. Actions against military personnel may include those recognized by the Manual for Courts-Martial (U.S.), 1969 (revised), paragraph 128c, or provided by regulation. Administrative action against civilian personnel may be pursued under U.S. Army civilian personnel regulations (AR 380-5, paragraph 14-101).

Evaluation Preparation: Setup: Provide the Soldier with appropriate guidance and material. Brief Soldier: Tell the Soldier to perform each step correctly and the material is burned or shredded.

Performance Measures	GO	NO-GO

1. Applied need-to-know limitations for access to classified material. (Refer to AR 380-5 [para 7-100] and TB 380-41.)
   a. Security clearance.
   b. Initial security briefing.
   c. Not solely by rank or position.
	___	___

2. Determined the classification of material to be destroyed.	___	___
3. Verified/prepared the DA Form 3964 or certificate of destruction.	___	___
4. Verified that the witness has the appropriate clearance.	___	___

5. Identified individual responsibilities and violations subject to sanctions. (Refer to AR 380-5 [para 14-101].)
   a. Violations.
   b. Sanctions.
	___	___

STP 31-18E34-SM-TG

Performance Measures	GO	NO-GO

6. Applied physical and administrative measures for storage and safekeeping of classified material. (Refer to AR 380-5 [paras 5-102, 5-104, and 5-202] and TB 380-41).
    a. Storage equipment.
    b. Recording storage facility data.
    c. Security checks.

7. Transported the material to be destroyed to the appropriate facility.

8. Applied control measures to prevent unauthorized entry and protect and safeguard classified material. (Refer to AR 380-5 [para 7-100] and (O)TB 380-41.)
    a. Need-to-know.
    b. Clearance status roster.
    c. Visitor register.
    d. Equipment screening.
    e. Visitor surveillance.

9. Assisted in the routine destruction of classified material. (Refer to AR 380-5 [paras 9-100 through 9-104].)
    a. Policy.
    b. Records.
    c. Methods.

10. Completed DA Form 3964 or certificate of destruction.

11. Forwarded DA Form 3964 or certificate of destruction to the security custodian, or filed as required.

Note: Specific extracts of the references are not included because of the quantity of material.

Evaluation Guidance: Score the Soldier GO if all performance measures are passed (P). Score the Soldier NO-GO if any performance measure is failed (F). If the Soldier fails any performance measure, show what was done wrong and how to do it correctly.

References

Required	Related
AR 380-5	
DA FORM 3964	
TB 380-41	

STP 31-18E34-SM-TG

Prepare Signal Annex of the Operation Order
113-611-6112

Conditions: In a field or garrison environment, given an incomplete unit operation order (OPORD) with mission statement, unit's signal operating instructions (SOI), and a map of the operational area.

Standards: Prepared a signal annex (Paragraph 5) by identifying the battalion's communications-electronics (CE) responsibility to the mission with supervisor approval in accordance with (IAW) associated references.

Performance Steps

1. Review supporting documents.
    a. Determine the mission.
    b. Determine the CE responsibility to the mission. (Review the unit's OPORD).
2. Prepare input for the applicable portion of the unit's SOI to be in effect. Prepare any special instructions relating to signal matters.
3. List the locations of the command posts (CPs).
    a. List CP location for the issuing unit.
    b. List CP locations for subordinate units.
    c. List CP location for the command group.
4. Determine if information on future locations of major headquarters (HQ) is needed. List all future signal system planning and location requirements as necessary.
5. Analyze the mission as it pertains to the CE responsibility.
    a. Ensure all input information to the CE of the Signal Annex is listed.
    b. Identify the CE responsibility to the mission.
6. Submit input information for supervisor's approval.

Evaluation Preparation: Provide the Soldier with all the material and equipment listed in the condition statement.

Performance Measures	GO	NO-GO
(Refer to unit's SOI and mission statement for performance measures 1 through 5.)		

1. Reviewed supporting documents.
    a. Determined the mission.
    b. Determined the CE responsibility to the mission. (Reviewed the unit's OPORD).
	——	——

2. Prepared input for the applicable portion of the unit's SOI to be in effect. Prepared any special instructions relating to signal matters.	——	——
Note: Refer to the map for performance measures 3 through 5.		

3. Listed the locations of the CPs.
    a. Listed CP location for the issuing unit.
    b. Listed CP locations for subordinate units.
    c. Listed CP location for the command group.
	——	——

4. Determined if information on future locations of major HQ are needed. List all future signal system planning and location requirements as necessary.	——	——

STP 31-18E34-SM-TG

Performance Measures   <u>GO</u>   <u>NO-GO</u>

   5. Analyzed the mission as it pertained to the CE responsibility.   ——   ——
      a. Ensured all input information to the CE of the Signal Annex was listed.
      b. Identified the CE responsibility to the mission.

   6. Submitted input information for supervisor's approval.   ——   ——

Evaluation Guidance: Score the Soldier GO if all performance measures are passed (P). Score the Soldier NO-GO if any performance measure is failed (F). If the Soldier fails any performance measure, show what was done wrong and how to do it correctly.

References
   Required   **Related**
   UNIT OPORD

STP 31-18E34-SM-TG

Implement Defense Information Operations
158-350-0001

Conditions: In a classroom, given the appropriate references and student handout.

Standards: Accurately:
(1) Define information operations (IO);
(2) Identify how the effects of defensive IO contribute to achieving information superiority;
(3) Identify three sources of threats to friendly information and information systems (INFOSYS);
(4) Identify three ways hostile intelligence threats collect information on friendly operations;
(5) Identify the steps of the operations security (OPSEC) process;
(6) Identify the three types of OPSEC indicators;
(7) Identify the purpose of the OPSEC survey;
(8) Identify three OPSEC measures to protect critical information;
(9) Identify three electronic protection procedures to protect radio communication;
(10) Identify three measures to protect information on the unit's Automated Information System (AIS).

Performance Steps

1. Define information operations.
2. Identify how the effects of defensive IO contribute to achieving information superiority.
3. Identify three sources of threats to friendly information and INFOSYS.
4. Identify three ways hostile intelligence threats collect information on friendly operations.
5. Identify the steps of the OPSEC process.
6. Identify the three types of OPSEC indicators.
7. Identify the purpose of the OPSEC survey.
8. Identify three OPSEC measures to protect critical information.
9. Identify three electronic protection procedures to protect radio communication.
10. Identify three measures to protect information on the unit's AIS.

Performance Measures	GO	NO-GO
1. Defined information operations.	——	——
2. Identified how the effects of defensive IO contribute to achieving information superiority.	——	——
3. Identified three sources of threats to friendly information and INFOSYS.	——	——
4. Identified three ways hostile intelligence threats collect information on friendly operations.	——	——
5. Identified the steps of the OPSEC process.	——	——
6. Identified the three types of OPSEC indicators.	——	——
7. Identified the purpose of the OPSEC survey.	——	——
8. Identified three OPSEC measures to protect critical information.	——	——

STP 31-18E34-SM-TG

Performance Measures   **GO**   **NO-GO**

  9. Identified three electronic protection procedures to protect radio communication.   —— ——

  10. Identified three measures to protect information on the unit's AIS.   —— ——

Evaluation Guidance: Score the Soldier GO if all performance measures are passed (P). Score the Soldier NO-GO if any performance measure is failed (F). If the Soldier fails any performance measure, show what was done wrong and how to do it correctly.

References
    Required   **Related**
    AR 530-1
    JP 3-13

Develop a Communications Plan
331-18E-3032

Conditions: Given a mission for a Special Forces operational detachment A (SFODA), operation order (OPORD) or operation plan (OPLAN), and unit signal operation instructions (SOI).

Standards: Develop a communications plan in accordance with (IAW) the performance measures.

Performance Steps

1. Review the OPORD or OPLAN.
    a. Determine the type of mission.
    b. Determine the number of communication nets required (internal and external).
    c. Review the SOI for completeness and clarity.
2. Determine how much and what type of communications equipment is needed and available.
    a. Ensure compatibility and frequency operating range of equipment to be used.
    b. Ensure proper secure equipment is on hand.
    c. Determine auxiliary (AUX) systems to be used.
    d. Determine how many batteries are needed for the different systems for entire length of mission.
3. Prepare diagrams, overlays, and schematics required for briefback.

Evaluation Preparation: Setup: Provide the Soldier with a mission for an SFODA, OPORD or OPLAN, and unit SOI.

Performance Measures	GO	NO-GO
1. Reviewed the OPORD or OPLAN.	——	——
2. Determined the type and amount of communications equipment needed.	——	——
3. Prepared diagrams, overlays, and schematics required for briefback.	——	——

Evaluation Guidance: Score the Soldier GO if all performance measures are passed (P). Score the Soldier NO-GO if any performance measure is failed (F). If the Soldier fails any performance measure, show what was done wrong and how to do it correctly.

References

Required	Related
	FM 24-1
	FM 5-0
	GTA 31-01-003

STP 31-18E34-SM-TG

Occupy a Transmission Site
331-18E-3058

Conditions: Given unit signal operation instructions (SOI), map of the area of operations (AO), lensatic compass, and protractor.

Standards: Select a Special Forces transmission site in accordance with (IAW) the training information outline.

Performance Steps

1. Identify the purpose of the transmission site.
    a. Ground-wave communications use line-of-sight (LOS) direct-wave or surface-wave transmission. Site must provide an unobstructed line-of-communications (LOC) path with no large terrain features between the transmitter and the receiver.
    b. Sky-wave communications use atmospheric propagation/transmission. The transmitter and receiver and site should meet the following requirements:
        (1) Vertical clearance of ridge lines in the direction of the transmission must be no more than one-half the required transmission angle.
        (2) The dimensional areas for the construction of the antenna must be sufficient for the height and the horizontal areas needed with the supporting elements in the correct location.

2. Conduct map reconnaissance to select tentative sites considering enemy positions, terrain, and any other operational considerations.

3. Evaluate tentative sites.
    a. Ensure security considerations are met.
        (1) Cover and concealment.
            (a) Equipment and personnel must be camouflaged to prevent detection from overhead, horizontal, and ground observation. Equipment must be packed for immediate evacuation.
            (b) Antenna must be erected in areas that provide cover from shrapnel and small-arms fire and concealment from observation. Antennas erected in fairly open terrain can be observed from long distances.
            (c) Depending on the location of the site, cover should be provided for protection against mortar, rocket, artillery, small-arms, and shrapnel damage.
        (2) Avenues of approach.
            (a) Vehicular avenues of approach should not only be limited to just established roads, but also should include any terrain that military vehicles can cross fast enough to prevent safe evacuation of the transmission site.
            (b) High-speed avenues of approach by foot (such as trails) should be avoided, especially in areas with heavy local populations.
            (c) Sites should not be located near clear areas that can be used for drop zones (DZs) and landing zones (LZs), especially if the enemy has the capability to conduct these operations.
        (3) Physical security.
        (4) Communication security (COMSEC).
    b. Ensure terrain considerations include—
        (1) Terrain that provides maximum masking of the signal in all but the desired direction of transmission.
        (2) Terrain that provides a good conductivity factor, especially for ground-wave sites.
            (a) The water table in the area should be confirmed, if possible, with transmission sites selected in areas of high water tables.
            (b) If poor conductivity exists, enhance this factor by using an artificial ground system.
            (c) Areas of high mineral deposits should be avoided because of the unpredictable nature of the reflected wave.

STP 31-18E34-SM-TG

Performance Steps

4. Select primary and alternate transmission sites.

5. Recover all communications equipment, such as antenna wire, transmission line, guy ropes, counterpoise, ground wire, batteries, radio components, and accessory equipment used.

6. Remove all signs of occupancy, such as trash, food wrappers, cigarette butts, human waste, crushed undergrowth, paths, and footprints.

7. Inspect the sterilized site to ensure nothing has been left behind and it has been returned to a natural-looking state.

8. Conceal tracks while exiting the site.

Evaluation Preparation: Setup: Provide the Soldier with a unit SOI, map of the AO, lensatic compass, and protractor.

Performance Measures	GO	NO-GO
1. Identified the purpose of the transmission site.	——	——
2. Conducted map reconnaissance to select tentative sites.	——	——
3. Evaluated tentative sites.	——	——
4. Selected primary and alternate transmission sites.	——	——
5. Recovered all communications equipment.	——	——
6. Removed all signs of occupancy.	——	——
7. Inspected the sterilized site.	——	——
8. Concealed tracks while exiting the site.	——	——
9. Assigned a surveillance team to watch the site for 48 hours.	——	——

Evaluation Guidance: Score the Soldier GO if all performance measures are passed (P). Score the Soldier NO-GO if any performance measure is failed (F). If the Soldier fails any performance measure, show what was done wrong and how to do it correctly.

STP 31-18E34-SM-TG

Compute Communications Equipment Electrical Requirements
331-18E-3076

Conditions: Given paper; pencil; calculator; FM 5-424, *Theater of Operations Electrical Systems*; and an itemized list of equipment requiring alternating current/direct current (AC/DC) power, in a field or garrison environment.

Standards: Compute the communications equipment electrical requirements within 30 minutes.

Performance Steps

1. Determine the communications equipment requiring electrical power.
2. Determine the required load for the communications equipment.
    a. Determine total wattage for all loads.
    b. Consider power factors of inductive loads.
3. Determine the demand load for the communications equipment facility.
    a. Multiply the connected load by the demand factor.
    b. Use Table 8-3, Demand Factors, found in FM 5-424.
4. Notify the engineer sergeant if the demand load exceeds the electrical supply.

Evaluation Preparation: Setup: Provide the Soldier with paper, pencil, calculator, FM 5-424, and an itemized list of equipment requiring AC/DC power to be used on deployment.

Performance Measures	GO	NO-GO
1. Determined the communications equipment requiring electrical power.	——	——
2. Determined the required load for the communications equipment.	——	——
3. Determined the demand load for the communications equipment facility.	——	——
4. Notified the engineer sergeant if the demand load exceeded the electrical supply.	——	——

Evaluation Guidance: Score the Soldier GO if all performance measures are passed (P). Score the Soldier NO-GO if any performance measure is failed (F). If the Soldier fails any performance measure, show what was done wrong and how to do it correctly.

References

Required	Related
FM 5-424	

STP 31-18E34-SM-TG

Manage Classified Material
331-18E-3077

Conditions: As a Special Forces communications sergeant in a field or garrison environment, given proper instruction, paper, pencil, and the requirement to manage classified material.

Standards: Manage the classified material in accordance with (IAW) Army Regulation (AR) 380-5, *Department of the Army Information Security Program*, and installation policies in order to prevent loss or compromise of classified material and equipment.

Performance Steps

1. Receive classified material.
    a. Examines material received.
    b. Receipts for material received.
    c. Records receipt of material.
    d. Secures and safeguards material.

2. Maintain classified material.
    a. Stores classified material.
    b. Inventories and accounts for classified material.
    c. Reviews and updates classified material.
    d. Requests required updated classified material.

3. Secure classified material recovered or discovered in an unsecured area.
    a. Secures classified material.
    b. Inventories classified material.
    c. Obtains statements of facts pertaining to the classified material.
    d. Notifies the communications security (COMSEC) account custodian of classified material recovered or discovered and circumstances.

4. Dispose of classified material.
    a. Transfers, issues, or destroys classified material.
    b. Records appropriate action.
    c. Reports transactions as required.

Evaluation Preparation: Setup: Provide the Soldier with proper instruction, paper, pencil, and the requirement to manage classified material.

Performance Measures	GO	NO-GO
1. Received classified material.	——	——
2. Maintained classified material.	——	——
3. Secured classified material recovered or discovered in an unsecured area.	——	——
4. Disposed of classified material.	——	——

Evaluation Guidance: Score the Soldier GO if all performance measures are passed (P). Score the Soldier NO-GO if any performance measure is failed (F). If the Soldier fails any performance measure, show what was done wrong and how to do it correctly.

References

Required	Related
	AR 380-5

STP 31-18E34-SM-TG

Subject Area 3: Communication Systems

Prepare Single-Channel Ground and Airborne Radio System (Manpack) for Operation
113-587-1064

Conditions: Given an operational single-channel ground and airborne radio system (SINCGARS) manpack radio with battery box CY-8346; battery BA-5590; antenna AS-3683; handset H-250; pack frame with straps; TM 11-5820-890-10-1, *Operator's Manual for SINCGARS Ground Combat Net Radio, ICOM Manpack Radio AN/PRC-119A*; and TM 11-5820-890-10-3, *Operator's Manual for SINCGARS Ground Combat Net Radio, Non-ICOM Manpack Radio AN/PRC-119*.

Standards: Standard is met when battery, fill battery, antenna and handset are installed and attached to pack frame in accordance with TM 11-5820-890-10-1.

Performance Steps

1. Install the battery.
   a. Install one battery used for the main power in the (SIP) radio (rechargeable BB-390 A/U battery, rechargeable BB-590/U battery, or nonrechargeable BB-5590/U (lithium) battery.
   b. Connect the battery case to the radio set. (Refer to TM 11-5820-890-10-8, *Operator's Manual for SINCGARS Ground Combat Net Radio, ICOM Manpack Radio AN/PRC-119A*, page 2-14.)
2. Assemble the radio and the pack frame. Assemble the pack and fasten the assembled radio into the carrying case. (Refer to TM 11-5820-890-10-8.)
3. Install the antenna. Connect the AS-3683/PRC to the radio. (Refer to TM 11-5820-890-10-8, page 5-3.)
4. Connect the handset H-250. Connect the handset H-250/U or handheld remote control radio device (HRCRD) (C-12493/U). (Refer to TM 11-5820-890-10-8, page 3-12.)
5. Preset function controls. (Refer to TM 11-5820-890-10-8, page 5-1.)

Evaluation Preparation: Setup: Provide the Soldier with SINCGARS manpack radio with battery box CY-8346, battery BA-5590, antenna, handset, pack frame, and appropriate TMs.

Performance Measures	GO	NO-GO
1. Installed batteries.	——	——
2. Installed antenna.	——	——
3. Connected handset H-250.	——	——
4. Assembled radio and pack frame.	——	——
5. Preset function controls.	——	——

Evaluation Guidance: Score the Soldier GO if all performance measures are passed (P). Score the Soldier NO-GO if any performance measure is failed (F). If the Soldier fails any performance measure, show what was done wrong and how to do it correctly.

References
Required	Related
TM 11-5820-890-10-1	
TM 11-5820-890-10-3	
TM 11-5820-890-10-8	

Install Single-Channel Ground and Airborne Radio Systems ICOM With or Without the AN/VIC-1 or AN/VIC-3
113-587-1067

Conditions: Given a designated vehicle; designated Single-Channel Ground and Airborne Radio Systems (SINCGARS) ICOM (with or without intercommunications set AN/VIC-1); installation kit(s) MK-2310 and MK-2314 (if radio set AN/VRC-92 is installed); automated network control device (ANCD) AN/CYZ-10 with fill, fill cable, and digital multimeter AN/PSM-45(*); tool kit TK-101/G; drill and drill bits from auto common No. 1; Technical Manual (TM) 11-5820-890-10-1, *Operator's Manual for SINCGARS Ground Combat Net Radio, ICOM Manpack Radio AN/PRC-119A*; TM 11-5820-890-20-1, *Unit Maintenance Manual for Ground ICOM Radio Set AN/PRC-119A*; TM 11-5820-890-20-2, *Unit Maintenance Manual for Ground ICOM Radio Set AN/PRC-119A*; TM 11-5830-340-12, *Operator's and Unit Organizational Maintenance Manual for Intercommunication Set AN/VIC-1(V)*; Service Bulletin (SB) 11-131-2, *Vehicular Radio Sets and Authorized Installations, Volume II: SINCGARS, FHMUX, and EPLRS*; and Department of the Army (DA) Form 5988-E (Equipment Inspection Maintenance Worksheet).

Standards: The standards are met when the radio set is mounted on its platform and test traffic is processed without error.

Performance Steps

1. Inventory complete radio system kit. (Refer to SB 11-131-2, TM 11-5820-890-10-1, TM 11-5820-890-20-1, and TM 11-5820-890-20-2.)
2. Install installation kit. (Refer to TM 11-5820-890-20-1, TM 11-5820-890-20-2, and TM 11-5830-340-12.)
3. Mount system components. (Refer to TM 11-5820-890-20-1 and TM 11-5820-890-20-2.)
4. Perform an operational check. (Pass test traffic.) (Refer to TM 11-5820-890-10-1.)
5. Establish maintenance records.

Evaluation Preparation: Setup: Provide the Soldier with designated vehicle, SINCGARS ICOM, installation kits, ANCD AN/CYZ-10, multimeter, TK-101/G, drill, appropriate TMs, SB, and DA form to mount radio set and run test traffic.

Performance Measures	GO	NO-GO
1. Inventoried complete radio system kit. (Refer to SB 11-131-2, TM 11-5820-890-10-1, TM 11-5820-890-20-1, and TM 11-5820-890-20-2.)	——	——
2. Installed installation kit. (Refer to TM 11-5820-890-20-1, TM 11-5820-890-20-2, and TM 11-5830-340-12.)	——	——
3. Mounted system components. (Refer to TM 11-5820-890-20-1 and TM 11-5820-890-20-2.)	——	——
4. Performed an operational check. (Passed test traffic.) (Refer to TM 11-5820-890-10-1.)	——	——
5. Established maintenance records.	——	——

STP 31-18E34-SM-TG

Evaluation Guidance: Score the Soldier GO if all performance measures are passed (P). Score the Soldier NO-GO if any performance measure is failed (F). If the Soldier fails any performance measure, show what was done wrong and how to do it correctly.

References
    Required   **Related**
    DA FORM 5988-E
    SB 11-131-2
    TM 11-5820-890-10-1
    TM 11-5820-890-20-1
    TM 11-5820-890-20-2
    TM 11-5830-340-12

STP 31-18E34-SM-TG

Construct Vertical Half-Rhombic Antenna
113-596-1052

Conditions: Given 200 feet of W-1 antenna wire, electrical tape, 400 to 600-ohm terminating resistor, insulators (or material to construct field-expedient insulators), radio set, knife, suspension line, measuring device, compass, signal operating instructions (SOI) with frequency and call signs, and a suitable training site.

Standards: Constructed a vertical half-rhombic antenna and it was operational.

Performance Steps

1. Construct antenna.
   Note: Counterpoise should be the length from leg closest to distant station to leg furthest from distant station.
   a. Measure 100 feet of antenna wire total length. (Add 12 inches to connect insulators.)
   b. Bend wire in half to find apex point for connection of insulator.
   c. Connect insulator to apex point.
   d. Connect halyard to insulator at apex point.
   e. Tape wire to insulator to prevent movement to apex point.

2. Erect antenna.
   a. Determine azimuth to distant station.
   b. Install antenna so apex is at least 30 feet high.
   c. Separate legs and equal distance on azimuth to distant station.
   d. Install counterpoise.
   e. Connect terminating resistor.
   Note: One end of resistor is connected to end of antenna; the other end is connected to counterpoise and closet to distant station.

3. Connect feeder line.
   a. Connect one wire to antenna and positive side to radio.
   b. Connect other wire from end of counterpoise to ground on radio.

4. Call distant station.

Performance Measures	GO	NO-GO
1. Constructed antenna.	___	___
2. Erected antenna.	___	___
3. Connected feeder line.	___	___
4. Called distant station.	___	___

Evaluation Guidance: Score the Soldier GO if all performance measures are passed (P). Score the Soldier NO-GO if any performance measure is failed (F). If the Soldier fails any performance measure, show what was done wrong and how to do it correctly.

References
   Required   Related
   UNIT SOI

STP 31-18E34-SM-TG

Construct a Long-Wire Antenna
113-596-1056

Conditions: Given a suitable area, paper, pencil, antenna wire, three insulators, a 50-foot guy rope, anchor stakes, hammer, knife, pliers, measuring tape, compass, frequency, a suitable radio, and an azimuth to orient the antenna.

Note: The AN/PRC-74 radio set is not a suitable radio due to the impedance mismatch.

Standards: Constructed and erected the antenna so it is within ±3 inches of the required length.

Performance Steps

1. Compute antenna length by using the following formula:
   $L = 492 (N - .05) \div$ Frequency (MHZ); L = length in feet; N = number of half wavelengths; (1 wavelength = 2 half wavelengths).
   Example: Build a 4-wavelength antenna for a frequency of 12 MHz.
   $L = 492 (N - .05) \div$ Frequency; $L = 492 (8 - .05) \div 12$; $L = 492 \times 7.95 \div 12$; $L = 326$ feet.

2. Assemble antenna.
   a. Measure antenna wire using computations from performance step 1.
   b. Attach insulators.

3. Install antenna.
   a. Using the compass, orient antenna to the direction of maximum desired radiation. (Long wires can be directional antennas.)
   b. Erect antenna.

Performance Measures

	GO	NO-GO
1. Computed antenna length by using the formula.	——	——
2. Assembled antenna.	——	——
3. Installed antenna.	——	——

Evaluation Guidance: Score the Soldier GO if all performance measures are passed (P). Score the Soldier NO-GO if any performance measure is failed (F). If the Soldier fails any performance measure, show what was done wrong and how to do it correctly.

Construct a Doublet Antenna
113-596-1070

Conditions: Given a requirement and a radio set; antenna group AN/GRA-50 (or sufficient W-1 antenna wire for the construction of the doublet antenna to the assigned frequency); compass (M-2 or equivalent); Technical Manual (TM) 11-5815-334-10, *Operator's Manual for Radio Teletypewriter Set, AN/GRC-122*; TM 11-5820-467-15, *Operator's, Organizational, Direct Support, General Support, and Depot Maintenance Manual for Antenna Group AN/GRA-50*; and wire cutter kit (TE-33 or equivalent).

Note: Supervision and assistance are available.

Standards: Properly cut doublet antenna frequency and erected broadside to the most distant station.

Performance Steps
Note: Performance step 3 is a team task.

1. Construct antenna using W-1 antenna wire (refer to TM 11-5815-334-10 and TM 11-5820-467-15); or construct antenna using antenna group AN/GRA-50 (refer to TM 11-5820-467-15).
    a. Use formula 468 ÷ Frequency = Length.
    b. Frequency____MHz.

Example: 468 ÷ 26.00 MHz = Length; 26x 18 feet = 468; 18 feet half wave or 9 feet quarter wave center fed.

2. Prepare mast AB-155(*)/U for erection. (Refer to TM 11-5815-334-10.)
3. Erect antenna. (Refer to TM 11-5815-334-10.)
    a. Antenna must be broadside to the most distant station. Determine azimuth by using compass.
    b. Connect antenna lead-in to radio set.

Performance Measures	**GO**	**NO-GO**
Note: Performance measure 3 is a team task.		

1. Constructed antenna using W-1 antenna wire (refer to TM 11-5815-334-10 and TM 11-5820-467-15); or constructed antenna using antenna group AN/GRA-50 (refer to TM 11-5820-467-15).
    a. Use formula 468 ÷ Frequency = length.
    b. Frequency _____ MHz.
Example: 468 ÷ 26.00 MHz = Length; 26 x 18 feet = 468; 18 feet half wave or 9 feet quarter wave center fed.
	——	——

2. Prepared mast AB-155(*)/U for erection. (Refer to TM 11-5815-334-10.)	——	——

3. Erected antenna. (Refer to TM 11-5815-334-10.)
    a. Antenna must be broadside to the most distant station. Determine azimuth by using compass.
    b. Connect antenna lead-in to radio set.
	——	——

Evaluation Guidance: Score the Soldier GO if all performance measures are passed (P). Score the Soldier NO-GO if any performance measure is failed (F). If the Soldier fails any performance measure, show what was done wrong and how to do it correctly.

References
Required	Related
TM 11-5815-334-10 TM 11-5820-467-15	TM 11-5820-520-10

STP 31-18E34-SM-TG

Operate Simple Key Loader AN/PYQ-10
113-609-2006

Conditions: Given a Simple Key Loader (SKL) AN/PYQ-10, appropriate radio set, platform, signal operating instructions (SOI), receive and transmit keys, and requirement to initialize the SKL to load a radio.

Standards: Soldier will initialize the SKL by performing required presets, managing the database, transferring a database, performing receive key procedures, viewing the menu and SOI menu, and loading the appropriate radio, and enabling the radio to both send and receive within 30 minutes.

Performance Steps

1. Perform presets.
    a. Use SKL buttons to calibrate stylus.
    b. Set auto power shutdown and backlight.
    c. Reset KOV-21 card.
    d. Create new single sign-on (SSO) password.
    e. Set KOV-21 card time and date.
    f. Check KOV-21 card time and date.
    g. View audit trail.
    h. View card status and audit trail free space.
    i. Clear audit trail.
    j. Create SKL user.
    k. Change user password.
    l. Launch the SKL UAS software.
    m. Activate night vision mode.
    n. Disable four-way button.
    o. Display summary status.
    p. Set key view option.
    q. Set profile mode.
    r. Set keys prompt.
    s. Set tree sort view.
    t. Set keys load.
    u. Enable audit warning message.

2. Manage database.
    a. Add platform.
    b. Add equipment.
    c. Create a traffic encryption key (TEK) key tag.
    d. Create a key encryption key (KEK) key tag.
    e. View key tag attributes.
    f. Assign key tag attributes.
    g. Assign equipment to platform.
    h. Filter platforms.
    i. Filter equipment.
    j. Set date load filter.
    k. Change key source.

3. Transfer database.
    a. Transfer ACES database to the SKL.
    b. Transfer data file from the jump drive to SKL.
    c. Transfer CT3 database to the SKL.
    d. Transfer database SKL to SKL.
    e. Transfer database single-channel ground and airborne radio system (SINCGARS) mode SKL to SKL.

8 February 2010

STP 31-18E34-SM-TG

Performance Steps
      f. Broadcast transmit windows.
      g. Broadcast receive windows.

  4. Receive key procedures.
      a. Receive key needed from KOK-22 (DS-101).
      b. Receive key from KOK-22 (DS-101).
      c. Receive key needed from KYK-13 (DS-102).

  5. View menu.
      a. Use find menu.
      b. Use matching keys menu.
      c. Delete expired keys.
      d. View and delete modern keys.

  6. SOI menu.
      a. Switch or select SOI edition.
      b. View SOI group information.
      c. View SOI TMPD information.
      d. View net information.
      e. View smoke and pyro information.

  7. Load radio.
      a. Load SINCGARS radio.
      b. Load selected location of the SINCGARS radio.
      c. Load a single location of the SINCGARS radio.
      d. Load single key to unknown device.

Performance Measures   <u>GO</u>   <u>NO-GO</u>

  1. Performed presets.   ____   ____
      a. Used SKL buttons to calibrate stylus.
      b. Set auto power shutdown and backlight.
      c. Reset KOV-21 card.
      d. Created new SSO password.
      e. Set KOV-21 card time and date.
      f. Checked KOV-21 card time and date.
      g. Viewed audit trail.
      h. Viewed card status and audit trail free space.
      i. Cleared audit trail.
      j. Created SKL user.
      k. Changed user password.
      l. Launched the SKL UAS software.
      m. Activated night vision mode.
      n. Disabled four-way button.
      o. Displayed summary status.
      p. Set key view option.
      q. Set profile mode.
      r. Set keys prompt.
      s. Set tree sort view.
      t. Set keys load.
      u. Enabled audit warning message.

  2. Managed database.   ____   ____
      a. Added platform.
      b. Added equipment.
      c. Created a TEK key tag.
      d. Created a KEK key tag.

8 February 2010   3-41

STP 31-18E34-SM-TG

Performance Measures	GO	NO-GO
e. Viewed key tag attributes.		
f. Assigned key tag attributes.		
g. Assigned equipment to platform.		
h. Filtered platforms.		
i. Filtered equipment.		
j. Set date load filter.		
k. Changed key source.		

3. Transferred database.   ____ ____
    a. Transferred ACES database to the SKL.
    b. Transferred data file from the jump drive to SKL.
    c. Transferred CT3 database to the SKL.
    d. Transferred database SKL to SKL.
    e. Transferred database SINCGARS mode SKL to SKL.
    f. Broadcasted transmit windows.
    g. Broadcasted receive windows.

4. Performed receive key procedures.   ____ ____
    a. Received key needed from KOK-22 (DS-101).
    b. Received key from KOK-22 (DS-101).
    c. Received key needed from KYK-13 (DS-102).

5. Viewed menu.   ____ ____
    a. Used find menu.
    b. Used matching keys menu.
    c. Deleted expired keys.
    d. Viewed and delete modern keys.

6. Viewed SOI menu.   ____ ____
    a. Switched or selected SOI edition.
    b. Viewed SOI group information.
    c. Viewed SOI TMPD information.
    d. Viewed net information.
    e. Viewed smoke and pyro information.

7. Loaded radio.   ____ ____
    a. Loaded SINCGARS radio.
    b. Loaded selected location of the SINCGARS radio.
    c. Loaded a single location of the SINCGARS radio.
    d. Loaded single key to unknown device.

References
    Required   **Related**
                                       TM 11-5810-410-13&P

STP 31-18E34-SM-TG

Operate Automated Net Control Device AN/CYZ-10
113-609-2053

Conditions: As a radio operator in a field environment, given automated net control device (ANCD) AN/CYZ-10 (C); Technical Bulletin (TB) 11-5820-890-12, *Operator and Unit Maintenance for AN/CYZ-10 Automated Net Control Device;* and Technical Manual (TM) 11-5820-890-10-8, *Operator's Manual for SINCGARS Ground Combat Net Radio, ICOM Manpack Radio AN/PRC-119A.* Given a requirement to operate the AN/CYZ-10.

Standards: Soldier performs in sequence the transfer of communications security (COMSEC) keys and signal operating instructions (SOI) information from ANCD to ANCD; loads SINCGARS radio with COMSEC variables using ANCD; and obtains SOI information from ANCD, correcting all errors within 15 minutes.

Performance Steps

1. Transfer COMSEC keys and SOI information from ANCD to ANCD.
    a. Turn on both ANCDs.
    b. Make main menu selection (ANCD).
    c. Make source ANCD menu selections.
    d. Make target ANCD Menu selections.
    e. Transfer data from ANCD to ANCD.
    f. Turn off/disconnect ANCDs.

2. Load radio from ANCD using Mode 2 fill.
    a. Turn radio and ANCD power On.
    b. Make main menu selection on ANCD.
    c. Make application menu selection on ANCD.
    d. Set controls of radio and connect to ANCD with fill cable.
    e. Transfer Mode 2 fill from ANCD to radio.
    f. Disconnect ANCD from radio and turn ANCD power OFF.

3. Obtain SOI information from ANCD.
    a. Turn ANCD power ON.
    b. Make main menu selection.
    c. Make SOI menu selection.
    d. Turn ANCD power OFF.

4. Perform preventive maintenance, checks, and services (PMCS) on ANCD.
    a. Make a visual Inspection of the ANCD.
    b. Check the battery.
    c. Check the fill port/crypto ignition key (CIK) port.
    d. Record entries on DA Form 2404 (Equipment Inspection and Maintenance Worksheet).

Performance Measures	GO	NO-GO
1. Transferred COMSEC data and SOI information from ANCD to ANCD.	——	——
2. Loaded Radio From ANCD using Mode 2 fill.	——	——
3. Obtained SOI information from the ANCD.	——	——
4. Performed PMCS on ANCD.	——	——

STP 31-18E34-SM-TG

Evaluation Guidance: Score the Soldier GO if all performance measures are passed (P). Score the Soldier NO-GO if any performance measure is failed (F). If the Soldier fails any performance measure, show what was done wrong and how to do it correctly.

References
    Required   **Related**
    TB 11-5820-890-12
    TM 11-5820-890-10-8

STP 31-18E34-SM-TG

Restore the Simple Key Loader AN/PYQ-10
113-609-4000

Conditions: Given an inoperable AN/PYQ-10; a computer; loaded simple key loader (SKL) AN/PYQ-10; fill cable; vendor's manuals; Technical Manual (TM) 11-5810-410-13&P, *Operator's and Field Maintenance Manual for Transfer Unit, Cryptographic Key AN/PYQ-10(C)*; and DA Form 5988-E (Equipment Inspection Maintenance Worksheet).

Standards: SKL has successfully transferred data or evacuated the defective line replaceable unit (LRU) to a higher maintenance level in accordance with (IAW) TM 11-5810-410-13&P and vendor's manuals.

Performance Steps

1. Verify reported malfunctions.
    a. Perform visual inspection.
    b. Review operator's actions.

2. Perform systematic troubleshooting procedures.
    a. Turn on and perform SKL software check.
    b. Login and launch user application software (UAS).
    c. Review database.

3. Take corrective actions.
    a. Perform self-test.
    b. Reload data using the SKL (source) or line control unit (LCU).
    c. Perform data transfer IAW vendor's manual.
    d. Evacuate the defective LRU to a higher maintenance level.
        (1) Prepare maintenance forms.
        (2) Process for a higher maintenance level.

Evaluation Preparation: Provide the Soldier with all the material and equipment listed in the condition statement.

Performance Measures	GO	NO-GO

1. Verified reported malfunctions.
    a. Performed visual inspection.
    b. Reviewed operator's actions.
	——	——

2. Performed systematic troubleshooting procedures.
    a. Turned on and perform SKL software check.
    b. Logged in and launched UAS.
    c. Reviewed database.
	——	——

3. Took corrective actions.
    a. Performed self-test.
    b. Reloaded data using the SKL (source) or LCU.
    c. Performed data transfer IAW vendor's manual.
    d. Evacuated the defective LRU to a higher maintenance level.
        (1) Prepared maintenance forms.
        (2) Processed for a higher maintenance level.
	——	——

8 February 2010

STP 31-18E34-SM-TG

Evaluation Guidance: Score the Soldier GO if all performance measures are passed (P). Score the Soldier NO-GO if any performance measure is failed (F). If the Soldier fails any performance measure, show what was done wrong and how to do it correctly.

References
    Required   **Related**
    DA FORM 5988-E
    TM 11-5810-410-13&P

STP 31-18E34-SM-TG

Install Radio Set AN/GRC-193A or Similar Radio Sets
113-620-1028

Conditions: Given an operational radio set AN/GRC-193A, operating frequency, a distant station, and Technical Manual (TM) 11-5820-924-13, *Operator's, Organizational, and Direct Support Maintenance Manual for Radio Set AN/GRC-193A.*

Standards: Performed all of the proper procedures to install the components of radio set AN/GRC-193A for voice operation in accordance with (IAW) TM 11-5820-924-13.

Performance Steps

1. Perform installation of the AM-6879 and the RT-1209.
    a. Loosen the wing nuts on the front of the mounting tray.
    b. Insert the back edge of the amplifier-converter assembly under the rear lip at the left side of the mounting tray.
    c. Position the front fasteners on MT-6232 to hold in and down on the front lip of the amplifier-converter.
    d. Tighten the wing nuts firmly.
    e. Secure the receiver/transmitter on the right side of the mounting tray by repeating steps a through d.

2. Perform installation of the AM-6545 and the CU-2064.
    a. Loosen the wing nuts at the right and left side of the mounting tray.
    b. Set the antenna coupler on the top left of the mounting tray.
    c. Position the fasteners of the mount to hold in and down on the side lip of the antenna coupler.
    d. Tighten the wing nuts properly.
    e. Set the power amplifier on the top right of the mounting tray. Position the side fasteners of the mount to hold in and down on the side lip of the power amplifier.
    f. Tighten the wing nuts firmly.

3. Perform cable connections.

> **CAUTION**
> Align all cable connectors before mating and fastening.

    a. Connect power control-radio frequency (RF) cable to receptacle J6 and connect the other end to receptacle J1.
    b. Connect the short audio cable to the bottom audio connector and the other end to the bottom audio connector.
    c. Install power control cable to cable receptacle J2 and the other end to receptacle J1.
    d. Connect RF cable to RF connector J1 and the other end to RF connector J4.
    e. Install power control cable to cable receptacle J2 and the other end to cable receptacle J5.

> **CAUTION**
> When making connection to the battery terminals of the vehicle, make sure connections are tight. Avoid accidental grounding of the positive terminal.

    f. Connect the direct current (DC) power cable to cable receptacle J3 and the other end to the battery of the vehicle. If using power supply PP-7333, connect it to J1 of the power supply.

STP 31-18E34-SM-TG

Performance Steps

    g. Connect RF cable to RF connector J5 and the other end to RF connector J7.

> **WARNING**
>
> There are 1600 volts present at antenna terminal J4 when using the WHIP antenna or at the J3 terminal when using the 50-ohm antenna. Do not remove the cables during operation. Extreme caution must be taken to ensure that these terminals are at least 6 inches away from nearby objects, such as cables, guy wires, brackets, or ground leads.

    h. Attach the whip antenna lead-in cable to whip antenna connector J4 and the other end to the whip antenna mounting base AB-652.
    i. When using doublet antenna, connect the 50-ohm coaxial cable to the 50-ohm connector J3 and the other end to the doublet antenna.
    j. Attach the ground strap to one of the terminals and the other end to the mount.
    k. Install the ground rod. Attach one end of the ground strap to one end of the terminal on front of antenna coupler and connect the other end to the ground rod.
    l. Connect the handset to top audio receptacle J1.

4. Install whip antenna.
    a. Assemble the whip antenna.
    b. Slide antenna cover over the assembled mast sections, tape the antenna tip assembly, and screw the antenna into the antenna mast base.
    c. Ensure antenna base is grounded securely.
    d. Tie the antenna down.
    e. Point the antenna in the direction of the distant station.

Performance Measures	GO	NO-GO
1. Performed installation of the AM-6879 and the RT-1209.	——	——
2. Performed installation of the AM-6545 and the CU-2064.	——	——
3. Performed cable connections.	——	——
4. Installed whip antenna.	——	——

Evaluation Guidance: Score the Soldier GO if all performance measures are passed (P). Score the Soldier NO-GO if any performance measure is failed (F). If the Soldier fails any performance measure, show what was done wrong and how to do it correctly.

References
    Required   **Related**
    TM 11-5820-924-13

STP 31-18E34-SM-TG

Install Improved High Frequency Radio Set AN/GRC-213 or a Similar System
113-620-1040

Conditions: Given a designated vehicle, improved high frequency radio (IHFR) set AN/GRC-213 (or a similar system); KY-99; appropriate installation kit (MK-2442, MK-2443, MK-2444, MK-2445, MK-2446, MK-2447, MK-2459, MK-2541, or MK-2544); digital multimeter AN/PSM-45(*); tool kit TK-101/G; drill and drill bits from auto common No. 1; and Technical Manual (TM) 11-2300-476-14&P, *Operator's Unit, Direct Support, and General Support Maintenance Manual for Installation Kits, Electronic Equipment MK-2442/GRC-213.*

Standards: The standard is met when the radio set is mounted on its platform and traffic is passed without error.

Performance Steps

1. Inventory installation kit. (Refer to TM 11-2300-476-14&P.)

2. Install installation kit. (Refer to TM 11-233-476-14&P.)
    a. Install mounting assembly.
    b. Install antenna bracket.
    c. Install antenna cables.
    d. Install power cables.
    e. Check for ground.

3. Configure system components. (Refer to TM 11-2300-476-14&P or the TM for the appropriate IHFR set to be installed.)

4. Perform operational check. (Refer to TM 11-2300-476-14&P or the TM of the appropriate IHFR set to be installed.)

5. Establish maintenance records.

Performance Measures	**GO**	**NO-GO**
1. Inventoried installation kit. (Refer to TM 11-2300-476-14&P.)	——	——
2. Installed installation kit. (Refer to TM 11-2300-476-14&P.) a. Installed mounting assembly. b. Installed antenna bracket. c. Installed antenna cables. d. Installed power cable. e. Checked for ground.	——	——
3. Configured system components. (Refer to TM 11-2300-476-14&P or the TM for the appropriate IHFR set to be installed.) a. Placed components in mount. b. Cabled components. c. Attached power cable. d. Attached antenna cables.	——	——
4. Performed operational check. (Refer to TM 11-2300-476-14&P or the TM of the appropriate IHFR set to be installed.)	——	——
5. Established maintenance records.	——	——

Evaluation Guidance: Score the Soldier GO if all performance measures are passed (P). Score the Soldier NO-GO if any performance measure is failed (F). If the Soldier fails any performance measure, show what was done wrong and how to do it correctly.

STP 31-18E34-SM-TG

References
    Required   **Related**
       TM 11-2300-476-14&P

STP 31-18E34-SM-TG

Employ International Maritime Satellite Terminal
331-18E-3019

Conditions: Given an MX2400T SATURN compact international maritime satellite (INMARSAT) terminal (complete), secure telephone unit (STU)-III, a requirement to transmit and receive a message, and applicable reference manuals.

Standards: Operate the INMARSAT terminal in accordance with (IAW) procedures outlined in the naval communications (NAVCOM) manual and the SATURN compact terminal.

Performance Steps

1. Install MX2400T INMARSAT terminal.

Note: Dual multi-tone frequency (DMTF) must be set on the STU-III telephone. Pick-up handset, press program (PGRM), then press *6; the display should read "DTMF HIGH"; if not, press 6 until it does. Hang up the handset.

    a. Install the terminal.
        (1) Open the zero suitcase.
        (2) Supply a 110- volt, alternating current (AC) power source.
        (3) Remove the keyboard from the lid of the suitcase.
        (4) Lift the flat display.
        (5) Connect the keyboard to the digital input (DIN) connector marked keyboard.
        (6) Turn the power on (takes 10 to 15 minutes).
    b. Install the antenna.
        (1) Lay the antenna transit case on its side and open.
        (2) Remove the tripod and deploy its adjustable legs.
        (3) Remove the "L" band electronics unit and insert the "L" band yoke into the top of the tripod.
        (4) Rotate the "L" band electronics unit 90 degrees to the ground with heat sink down.
        (5) Hand-tighten the blue locking knobs.
        (6) Remove the two halves of the dish from the transit case. Place the half with the Magnavox logo on the top of the "L" band electronics unit, slide it against the keyway with the logo on the same side as the "S" meter.
        (7) Secure the dish to the "L" band with two quarter-turn CAM-LOC fasteners.
        (8) Place side two (other half of the dish) on the "L" band electronics unit and press against side one so the indexing pins line up and fasten with the two CAM-LOC fasteners.
        (9) Fasten the four latches on the face of the disk.
        (10) Remove the horn assembly and press its connector side down in the center of the dish.
        (11) Secure in place with two threaded retainers on the support ribs on the back of the dish.
        (12) Aim the dish at the horizon with the meter on top.
        (13) Tighten the quarter-turn fasteners on the support ribs on the back of the dish.
        (14) Remove the antenna power supply and place it on the tripod cross-ribs.
        (15) Remove the 50-foot antenna control cable from the transit case and connect the 16-pin and terminal node controller (TNC) connector to the back of the "L" band electronics unit.
        (16) Secure the connector by rotating the outer ring clockwise until tightened.
        (17) Connect the other end to the matching connectors.

Note: DO NOT wrap the cables under the cross-ribs or around the legs of the tripod.

        (18) Remove the 50-foot cable from the transit case and connect the 9-pin connector and TNC connectors at the opposite end of the electronics console connectors marked ANTENNA (ANT).

STP 31-18E34-SM-TG

Performance Steps

      (19) Turn on the switch marked ANT (AC).
      (20) Place the compass on the back of the "L" band electronics unit.
    c. Initialize the MX2400T INMARSAT terminal.
      (1) Press F1 (screen appears).

Note:
1. When screen 1 is displayed the cursor will be over the letters "N" or "S."
2. Use the right arrow key to move to the value you want to change.
3. To change, use the up arrow key for higher and the down arrow key to lower values.
4. Shift arrow key allows changes in increments of 10. PgUp goes to highest value possible and PgDn goes to the lowest value possible.
5. If the F6 (SAVE) key has not been pressed, the F10 key will cancel the change.

      (2) Choose one of the following:
          (a) POSITION. Entries for latitude/longitude.

Note: If you know the magnetic variation, go to screen 5 and enter it under the antenna orientation.

          (b) REGION. Select satellite for the region of operation.
          (c) COAST EARTH STATION (CES). Shows the CES selected.
          (d) DAY/TIME. Enter current date and time.

Note: MESSAGE BOX in lower right hand screen displays messages from the central processing unit (CPU). Values that have been changed flash on and off. Press the F6 key to save. Press the F8 key to aim.

    d. Aim the antenna.
      (1) Press the F8 in screen 1. The MX2400T will compute the azimuth and elevation. They will be displayed on screen 4. To get to screen 4, press SHIFT/F4.
      (2) Screen 4, marked "antenna column one," is the azimuth relative to north, and column two is the elevation.
2. Operate the MX2400T INMARSAT terminal.
    a. Place a call.
      (1) Lift the handset, listen for a dial tone.
      (2) Dial subscriber's number (for example, 00 1 910 432 3633#).
         - 00 is for auto service.
         - 1 is the country code.
         - 910 is the area code.
         - 432 3633 is the subscriber's number.
         - # is the ending sign.
      (3) The connection is complete when the party answers go to secure on the STU-III telephone.
    b. Receive a call.
      (1) Lift the receiver and go secure on the STU-III.
      (2) Local calls: 1 ring every 6 seconds.
      (3) Voice call from CES: 2 rings every 6 seconds.
3. Install the SATURN compact INMARSAT terminal.
    a. Install the antenna.
      (1) Fasten the antenna rod to the terminal container.
      (2) Mount the antenna pivot on the top of the rod.
      (3) Mount the mid-dish section and feeder.
      (4) Connect the coax to the antenna and to the equipment.

STP 31-18E34-SM-TG

Performance Steps
        (5) Mount the dish side sections.
        (6) Mount the feeder reflector.
    b. Install the terminal.
        (1) Supply an AC power source (110 or 220 volts, direct current [VDC]).
        (2) Connect the cable from the main ID on the box to the phone.
    c. Initialize the terminal.
        (1) Turn on the main power. (NOTE: Terminal takes about 5 minutes to warm up.)
        (2) Terminal will run through the initializing sequence.
        (3) Press POINT and adjust the antenna orientation for maximum signal reading on the display.
        (4) Once the terminal has achieved a maximum signal reading, press OPERATE.

4. Place a call using the SATURN compact INMARSAT terminal.
    a. Lift the handset; listen for the dial tone.
    b. Dial the 4 digit code 0111.
    c. Listen in the handset for a 1.5 second tone.
    d. Dial subscriber's number (for example, 00 1 910 432 3633).
        - 00 is for auto service.
        - 1 is the country code.
        - 910 is the area code.
        - 432 3633 is the subscriber's number.
        - # is the ending sign.

5. Receive a call using the SATURN compact INMARSAT terminal.
    a. Lift the receiver and go secure on the STU-III.
    b. Local calls: 1 ring every 6 seconds.
    c. Voice calls from CES: 2 rings every 6 seconds.

Evaluation Preparation: Setup: Provide the Soldier with an MX2400T SATURN compact INMARSAT terminal (complete), STU-III, a requirement to transmit and receive a message, and applicable reference manuals.

Performance Measures	GO	NO-GO
1. Installed the MX2400T INMARSAT terminal.	——	——
2. Installed the antenna. (Note: Ensure the blue knobs are ONLY hand-tightened.)	——	——
3. Initialized the MX2400T INMARSAT terminal.	——	——
4. Placed a call using the INMARSAT.	——	——
5. Received an incoming call on the INMARSAT.	——	——

Evaluation Guidance: Score the Soldier GO if all performance measures are passed (P). Score the Soldier NO-GO if any performance measure is failed (F). If the Soldier fails any performance measure, show what was done wrong and how to do it correctly.

References
    Required   **Related**
                                                  LSS-94429
                                                  TT-98-107770B

STP 31-18E34-SM-TG

Operate the AN/PSC-5 in Satellite Communications Mode
331-18E-3024

Conditions: Given an AN/PSC-5 radio set (complete), selected frequencies, two BA5590 or BB490 batteries, and Technical Manual (TM) 11-5820-1130-12&P, *Operator's and Unit Maintenance Manual for Radio Set AN/PSC-5.*

Standards: Operate the AN/PSC-5 radio set in the satellite communications (SATCOM) mode in accordance with (IAW) procedures outlined in TM 11-5820-1130-12&P.

Performance Steps

1. Operate Radio set in SATCOM mode.

> **WARNING**
>
> SATCOM antennas concentrate transmitter signals into beams of high-energy electromagnetic radiation. DO NOT STAND in front of the satellite antenna or touch it at any time when transmitting. A distance of 10 inches should be maintained from the front of the antenna at all times to avoid partial body exposure which would exceed the applicable permissible limits. Avoid physical contact with any bare-metal wire or antenna surface because it could result in a radio frequency (RF) shock or burn.

   a. Point antenna in the direction of the satellite. Note: In plain text (PT) mode, only data operation is available.
   b. Set mode switch to PT or cipher text (CT) as required.
   c. Adjust dimness (DIM) control for desired display brightness.
   d. Press NEXT/PREV key to move cursor to mode field.
   e. Press arrow keys until SATCOM is displayed.
   f. Press enter (ENT) key.
   g. Select desired operating preset using keypad number key.
   h. Press ENT key.
   i. Select desired traffic encryption key (TEK) # (1-5) if in the CT aide using the keypad number key.

> **WARNING**
>
> Volume levels at the handset/headset/earphone/loudspeaker must be adjusted to the minimum levels required for operations. The volume control should be adjusted from the minimum position up to the comfortable level. Prolonged excessive volume will lead to hearing loss.

   j. Adjust VOLUME control as desired.
   k. Observe field strength indication on the display.
   l. Position satellite antenna for maximum field strength indication.

2. Operate radio set for voice operation.

STP 31-18E34-SM-TG

Performance Steps
    a. To transmit, press and hold push-to-talk (PTT) switch.
    b. Observe that transmit indication and signal strength are shown on display during transmission.

3. Operate radio set to receive.
    a. Release PTT switch.
    b. Listen to the handset ear piece.
    c. Observe that receive indication and signal strength is shown on display during reception.
    d. Rotate SQUELCH to control to eliminate background noise.

4. Operate radio set for data operations.
    a. Send message by keying data devices.
    b. Observe transmit indication on receiver/transmitter (R/T).
    c. Receive message on data devices.
    d. Observe receive indication on R/T.

Performance Measures   **GO**   **NO-GO**

1. Operated radio set in SATCOM mode.   ___   ___
    a. Pointed antenna in the direction of the satellite.
       Note: PT mode, only data operation is available.
    b. Set mode switch to PT or CT as required.
    c. Adjusted DIM control for desired display brightness.
    d. Pressed NEXT/PREV key to move cursor to mode field.
    e. Pressed arrow keys until SATCOM is displayed.
    f. Pressed ENT key.
    g. Selected desired operating preset using keypad number key.
    h. Pressed ENT key.
    i. Selected desired TEK # (1-5) if in the CT aide using the keypad number key.
    j. Adjusted VOLUME control as desired.
    k. Observed field strength indication on the display.
    l. Positioned satellite antenna for maximum field strength indication.

2. Operated radio set for voice operation.   ___   ___
    a. Transmitted by pressing and holding PTT switch.
    b. Observed that transmit indication and signal strength are shown on display during transmission.

3. Operated radio set to receive.   ___   ___
    a. Released PTT switch.
    b. Listened to the handset ear piece.
    c. Observed that receive indication and signal strength is shown on display during reception.
    d. Rotated SQUELCH to control to eliminate background noise.

4. Operated radio set for data operations.   ___   ___
    a. Sent message by keying data devices.
    b. Observed transmit indication on R/T.
    c. Received message on data devices.
    d. Observed receive indication on R/T.

Evaluation Guidance: Score the Soldier GO if all performance measures are passed (P). Score the Soldier NO-GO if any performance measure is failed (F). If the Soldier fails any performance measure, show what was done wrong and how to do it correctly.

References
    Required   **Related**
    TM 11-5820-1130-12&P

STP 31-18E34-SM-TG

Operate the AN/PSC-5C/D in Demand Assignment Multiple Access Mode
331-18E-3025

Conditions: Given an AN/PSC-5C/D radio set (complete), selected frequencies, two each BA5590 or BB490 batteries, Field Manual (FM) 6-02.53, and Technical Manual (TM) 11-5820-1130-12&P.

Standards: Operate the AN/PSC-5C/D radio set in the demand assigned multiple access (DAMA) mode in accordance with (IAW) procedures outlined in the TM 11-5820-1130-12&P.

Performance Steps

1. Operate radio set in DAMA mode.
   a. DAMA installation.
   (1) Point antenna in general direction of satellite.
   (2) Set mode switch to plain text (PT) or cipher text (CT) as required.
   (3) Adjust dimness (DIM) control for desired display brightness.
   (4) Press next previous (NEXT/PREV) keys to select CURRENT MODE MENU.
   (5) Press arrow keys to select DAMA.
   (6) Press enter (ENT) key.
   (7) Select desired operating preset (01-30) using keypad number keys.
   (8) Press ENT key.
   (9) In CT, select desired TEK # (1-19) using desired number key.

Note: The preset number (P#) field indicates the parameters displayed are stored as a preset. The P will change to a M (manual) if parameters are changed. Changes to the DAMA CURRENT MODE menu will not be enabled until the DAMA Operations menu is accessed. Until then, the terminal will continue to operate in the original settings. Also, if you escape from the CURRENT MODE menu after making the changes, and then return, the changes will not be enabled and the display will revert back to the original settings.

   (10) Press enter (ENT) key.
   (11) Observe field strength indication on display.
   (12) Position satellite antenna for maximum field strength indication.
   (13) If operational changes to the current mode are required, perform the following steps:
      (a) Press NEXT/PREV keys to move cursor to desired data field on CURRENT MODE menu.
      (b) Press arrow keys or number keys for desired changes.
      (c) Press ENT key.

WARNING

Satellite communications (SATCOM) antennas concentrate transmitter signals into beams of high energy electromagnetic radiation. DO NOT STAND in front of the satellite antenna or touch it at any time when transmitting. A distance of 10 inches should be maintained from the front of the antenna at all times to avoid partial body exposure which could exceed the applicable permissible limits. Avoid physical contact with any bare-metal wire or antenna surface because it could result in a radio frequency (RF) shock or burn.

---

   (14) Press NEXT/PREV to select DAMA Operations on the CURRENT MODE Menu.
   (15) Press ENT key.
   (16) Check menu settings and change as required.

STP 31-18E34-SM-TG

Performance Steps
    (17) Press NEXT to move cursor to "start DAMA for #####" (your address), press ENT.

  2. Operate radio set in 5 kHz DAMA mode.
Note: Upon entering the 5 kHz DAMA mode, the terminal is attempting to acquire satellite down link. The terminal will then range to satellite. This process may take up to 5 minutes. From the NETWORK 5kHz menu, observe for the following indications at the "Net" field.

    a. Idle—Terminal in idle state. No action is required.
    b. Acquiring—Downlink acquisition is process. No action is required.
    c. Range—5 kHz ranging in process. No action is required.
    d. Log in—5 kHz network acquired. Begin log in (go to Step 2a).
        (1) Press ENT key, when "Log" in is shown at the "Net" field.
        (2) Press arrow keys to select "Preassigned" or "Over the Air."
        (3) Press ENT key.
        (4) Press NEXT/PREV keys to move cursor to "Prec" field.
        (5) Press arrow keys to select the maximum terminal precedence for the preassigned log in.
        (6) Press ENT key.
        (7) With cursor on (SEND), Press ENT key.
        (8) Perform any of the following additional 5 kHz DAMA task as required:
            (a) Circuit service set up.
            (b) Dedicated service set up.
            (c) Message service set up.
            (d) Check service set up.
            (e) Check network state.
            (f) Check status messages.
            (g) Contention ranging.
            (h) Service teardown.
            (i) Log out.
            (j) Modify current mode.

  3. Operate radio set in 25 kHz automated control (AC) DAMA mode.
    a. From the DAMA Operations 25 kHz AC menu, observe for the following indications at the "Net" field:
        (1) Idle—Terminal in idle state. No action required.
        (2) Acquiring—Downlink acquisition is in process. No action is required.
        (3) Range—25 kHz ranging in process. No action is required.
        (4) Connected—Ready for activity (Go to step b.)
    b. Press ENT key.
    c. With cursor on SEND, press ENT key.
    d. At NETWORK 25kHz menu, press hot key #1 to select service setup.
    e. Press arrow keys to select the precedence of the service connection.
    f. Press ENT key.
    g. Press keypad number keys to enter party/parties to be called.
    h. Press ENT key.
    i. Perform the following steps to enter time a circuit connection is needed:
        (1) Use keypad number keys to enter 00 to 59 off time "##".
        (2) Press ENT key.
        (3) Press arrow keys to select sec, min, hrs, or day.
        (4) Press ENT key.
        (5) To select indefinite time, press NEXT to bypass "##".
        (6) Press arrow keys to select ind (indefinite).
        (7) Press ENT key.
    j. With cursor on SEND, press ENT key. Note: Display returns to the DAMA Operations menu.

STP 31-18E34-SM-TG

Performance Steps
    k. Observe for the following indication at the connections state:
        (1) Idle--Terminal in idle state. No action is required.
        (2) Pend--Request for connect started. No action is required.
        (3) Disc--Disconnecting (by controller) continue with communications.
        (4) Conn--Receive (RX)/transmit (TX) service connected for half duplex. Begin communications.
        (5) Conn RX--Service connected for receive only. Begin communications.
    l. For voice operation, perform the following steps:
        (1) To transmit, press and hold push-to-talk (PTT) switch and listen to the handset earpiece.
        (2) To receive, release PTT switch and listen to a handset earpiece.
    m. For data operation, perform the following steps:
        (1) Connect data device to radio set.
        (2) Send message by keying data device.
        (3) Ensure received messages are automatically output to data device.
    n. Perform any of the following additional 25 kHz AC DAMA tasks as required:
        (1) Checking service setup.
        (2) Checking network state.
        (3) Checking status messages.
        (4) Data transfer.
        (5) Link testing.
        (6) Paging.
        (7) Responding to information requests.
        (8) Service teardown.
        (9) Sending out-of-service message.
        (10) Dedicated channel operational.
        (11) Modify current mode.
4. Operate radio set in 25 kHz directed control (DC) DAMA mode.
    a. From the DAMA operations 25 kHz DC menu, observe for the following indications at the "Net" field:
        (1) Modify current mode.
        (2) Acquiring—Downlink acquisition is in process. No action is required.
        (3) Range—25 kHz ranging is in process. No action is required.
        (4) Connected—Terminal is ready for activity. (Go to step b.)
    b. At DAMA Operations 25 kHz menu, press hot key #1 to select service set-up.
    c. Press keypad number keys to select the circuit to be used.
    d. Press ENT key.
    e. Observe the third line of the display for the following indications:
        (1) Idle—Terminal in idle state. No action is required.
        (2) Conn-RX/TX—Service connected for half duplex. Begin communications.
        (3) Conn-RX—Service connected for receive only. Begin communications.
    f. For voice operation, perform the following steps:
        (1) To transmit, press hold PTT switch and listen to the handset earpiece.
        (2) To receive, release PTT switch and listen to handset earpiece.
    g. For data operation, perform the following:
        (1) Connect data device to radio set.
        (2) Send message by keying data device.
        (3) Ensure received messages are automatically output to data devices.
    h. Perform any of the following additional 25kHz DC DAMA tasks as required:
        (1) Checking service state.
        (2) Checking network state.
        (3) Checking status messages.
        (4) Data transfer.
        (5) Link testing.
        (6) Paging.

STP 31-18E34-SM-TG

Performance Steps

    (7) Responding to information requests.
    (8) Service teardown.
    (9) Modifying current mode.

Performance Measures	GO	NO-GO
1. Operated radio set in DAMA mode.	——	——
2. Operated radio set in 5 kHz DAMA mode.	——	——
3. Operated radio set in 25 kHz AC DAMA mode.	——	——
4. Operated radio set in 25 kHz DC DAMA mode.	——	——

Evaluation Guidance: Score the Soldier GO if all performance measures are passed (P). Score the Soldier NO-GO if any performance measure is failed (F). If the Soldier fails any performance measure, show what was done wrong and how to do it correctly.

References
    Required   **Related**
    TM 11-5820-1130-12&P
    FM 6-02.53

STP 31-18E34-SM-TG

Update Crypto Fill Keys on the AN/PSC-5
331-18E-3027

Conditions: Given an AN/PSC-5 radio set (complete), selected frequencies, two each BA5590 or BB490 batteries, and Technical Manual (TM) 11-5820-1130-12&P, *Operator's and Unit Maintenance Manual for Radio Set AN/PSC-5.*

Standards: Update the crypto fills on the AN/PSC-5 radio set in accordance with (IAW) procedures outlined in TM 11-5820-1130-12&P.

Performance Steps
Perform key update procedures.

Note: When in rotations mode, switch to update (UPD). (Do not to go past the UPD position to the Z position. This will zero the radio set.) The update function is irreversible. If you are requested to update more than once, observe that the update count processes before updating the second or additional times.

1. Set mode switch to UPD position.

Note: The display menu shows the communications security "(COMSEC) KEY UPDATE."

2. Select COMSEC key (1-5) to be updated using keypad number keys.
3. Observe key (1-5) to be updated using keypad number keys.
4. Press enter (ENT) key.

Note: Display shows the message.

5. Repeat steps b through e as required for any additional key updates.
6. Deselect UPD position with mode switch.

Performance Measures	GO	NO-GO
Performed key update procedures.		
1. Set mode switch to UPD position.	——	——
2. Selected COMSEC key (1-5) to be updated using keypad number keys.	——	——
3. Observed key (1-5) to be updated using keypad number keys.	——	——
4. Pressed ENT key.	——	——
5. Repeated steps 2 through 5 as required for any additional key updates.	——	——
6. Deselected UPD position with mode switch.	——	——

Evaluation Guidance: Score the Soldier GO if all performance measures are passed (P). Score the Soldier NO-GO if any performance measure is failed (F). If the Soldier fails any performance measure, show what was done wrong and how to do it correctly.

References
    Required   Related
    TM 11-5820-1130-12&P

STP 31-18E34-SM-TG

Perform Unit-Level Preventative Maintenance Checks and Services on Communications Equipment
331-18E-3028

Conditions: Given communications equipment, applicable technical manual (TM) for equipment; Department of the Army (DA) Form 2404 (Equipment Inspection and Maintenance Worksheet), DA Form 2407 (Maintenance Request), DA Form 2407-1 (Maintenance Request Continuation Sheet), and a pencil or pen.

Standards: Conduct preventative maintenance, checks, and services (PMCS) in accordance with (IAW) the TM. Note all deficiencies or services on appropriate forms.

Performance Steps

1. Perform the PMCS IAW the appropriate TM for the equipment.
2. Prepare DA Form 2404 (paragraph 3-4a to 3-4c).
3. Prepare DA Form 2407 (paragraph 3-6).
    a. Request support maintenance (paragraph 3-7b[2]).
    b. Report on accomplishment of a modification work order (paragraph 3-8a to 3-8c).
    c. Record work accomplished at support level (paragraph 3-6b[3][a]).
4. Prepare DA Form 2407-1 (paragraph 3-6).
5. Make proper disposition of DA Forms 2404, 2407, and 2407-1.
    a. DA Form 2404 disposition (paragraph 3-4d).
    b. DA Form 2407 disposition (paragraph 3-7d and 3-8d).
    c. DA Form 2407-1 disposition (paragraph 3-7d and 3-8d).

Evaluation Preparation: Setup: Provide the Soldier with communications equipment; applicable TM for equipment; DA Forms 2404, 2407, 2407-1; and a pencil or pen.

Performance Measures	GO	NO-GO
1. Performed the PMCS IAW the appropriate TM for the equipment.	——	——
2. Prepared DA Form 2404 (paragraph 3-4a to 3-4c).	——	——
3. Prepared DA Form 2407 (paragraph 3-6).	——	——
4. Prepared DA Form 2407-1 (paragraph 3-6).	——	——
5. Made proper disposition of DA Forms 2404, 2407, and 2407-1.	——	——

Evaluation Guidance: Score the Soldier GO if all performance measures are passed (P). Score the Soldier NO-GO if any performance measure is failed (F). If the Soldier fails any performance measure, show what was done wrong and how to do it correctly.

References
Required
DA FORM 2404
DA FORM 2407
DA FORM 2407-1

Related

STP 31-18E34-SM-TG

Use a Multimeter to Perform a Continuity Check and Voltage Check
331-18E-3029

Conditions: Given a multimeter, antenna wire, and a battery.

Standards: Set up a multimeter for operation, perform a continuity check on the antenna wire, and perform a voltage check on the battery within 10 minutes.

Performance Steps

1. Set up a multimeter for operation.
    a. Turn on the multimeter selector switch to the lowest ohm setting.
    b. Connect the red test lead to the positive jack.
    c. Connect the black test lead to the negative jack.
    d. Hold the end of black test lead and the end of red test lead together.
    e. Adjust the multimeter for a 0-ohm reading.
        (1) Observe the multimeter scale for 0 ohms.
        (2) Adjust the ohm knob until the scale reads 0 ohms.

2. Perform a continuity check on the antenna wire.
    a. Touch the red test lead to one end of the antenna wire, and, at the same time, touch the black test lead to the opposite end of the antenna wire.
    b. Observe that meter reading is 0 ohms.

3. Perform voltage check on the battery.
    a. Turn multimeter selector switch to direct current (DC) range and set the correct voltage setting (next highest setting above the voltage to be tested).
    b. Apply the positive and negative leads to the positive and negative terminals or the battery.
    c. Observe the meter for voltage reading.

Evaluation Preparation: Setup: Provide the Soldier with a multimeter, antenna wire, and a battery.

Performance Measures	GO	NO-GO
1. Set up a multimeter for operation.	——	——
2. Performed continuity check on the antenna wire.	——	——
3. Performed voltage check on the battery.	——	——

Evaluation Guidance: Score the Soldier GO if all performance measures are passed (P). Score the Soldier NO-GO if any performance measure is failed (F). If the Soldier fails any performance measure, show what was done wrong and how to do it correctly.

References

Required	Related
	TM 11-6625-203-12

STP 31-18E34-SM-TG

Construct a Clandestine Antenna
331-18E-3042

Conditions: Given a high-frequency (HF) radio set, unit signal operation instructions (SOI), installation sites (indoor and outdoor, urban and rural), designated frequency, pencil and paper, lensatic compass, expedient insulator material, electrical tape, knife, and pliers.

Standards: Construct and install an operational indoor and outdoor clandestine antenna (urban or rural) within 1 hour for each antenna in accordance with (IAW) the performance measures.

Performance Steps

1. Construct a clandestine indoor antenna (urban or rural).
    a. Select the most suitable indoor location at the site for the required transmission.
        (1) Select a building in a sector of the city towards the receiving station so transmitted signal will not be distorted or attenuated by other structures. Avoid industrial areas and areas where buildings are made of reinforced steel (skyscrapers).
        (2) Check the area around the building location for television and radio reception antennas to determine the possibility of interference with local reception of television and radio signals. Avoid areas with many rooftop antennas, high-tension power lines, and heavy motor vehicle traffic. To detect such sources of interference, use a small amplitude-modulated (AM) transistor radio (with an earphone) set to the highest frequency.
        (3) Select as high a floor as possible within the chosen building so other close buildings do not interfere with transmitted signal. Install antenna on the side of the building in the direction of transmission if possible.
        (4) Check local available power for adequacy of voltage, current, and frequency requirements. Power sources must be sufficient enough to preclude fluctuation in other areas of the neighborhood or building when transmitting.
        (5) Check walls and ceiling of the room for reinforcing bars or beams, metal lathe plaster, or any substance that conducts electricity. Hold the compass close to the walls and ceiling to detect metallic objects. This procedure, however, will not detect nonmagnetic metal, so a thorough inspection must be made.
        (6) Use the radio receiver to check for local reception interference from faulty wiring, fluorescent lights, and other sources.
        (7) Determine if the noise created by the radio station will be detected by nearby rooms or passersby.
        (8) Determine the bearing to the receiving station in relation to the layout of the room.
        (9) Check the room (moldings, baseboards, and window frames) for means of concealment and support for antenna and transmission lines.
    b. Select a suitable antenna for the indoor site and required transmission.
        (1) Determine the length of antenna for the frequency given.
            (a) Figures 1 and 2 reflect the formulas used to determine the length of the antenna in feet according to the type of antenna constructed.
            (b) Because the loop antenna must be square, divide the length by four to determine the size of each leg of the square.
        (2) Use other antenna configurations if there is enough space within the indoor location.
    c. Assemble components.
        (1) Measure and cut antenna wire to a tolerance of 2 inches.
        (2) Attach loop to wall resulting in closest azimuth.
    d. Connect the antenna to radio set.
    e. Tune radio. (Radio should tune in 3 to 12 seconds.)

8 February 2010

Performance Steps

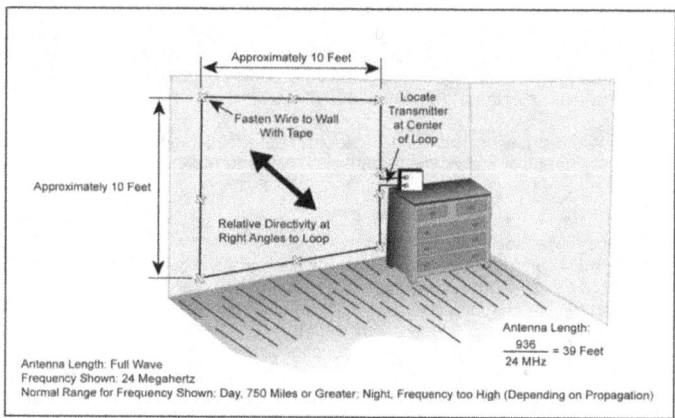

Figure 1. Full-wave square loop antenna

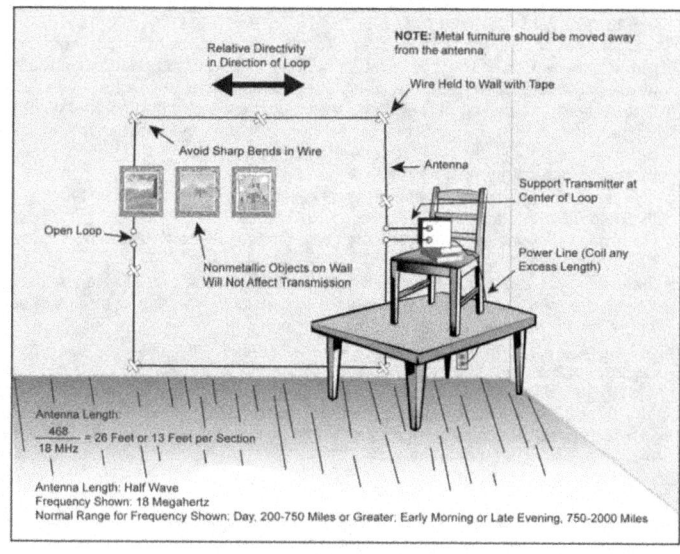

Figure 2. Half-wave square loop antenna

STP 31-18E34-SM-TG

Performance Steps

2. Construct an outdoor clandestine antenna (urban or rural).
    a. Determine if the radio station is to be permanent and operated from inside a structure or if the station is to be operated from the antenna site.
        (1) If the radio station is to be operated inside a structure, select a location as close as possible to the structure and one in which the transmission line will not be discovered during operation by passersby or damaged by vehicular or foot traffic.
        (2) Check area around location for television and radio reception antennas to determine the possibility of interference with local reception of television and radio signals. Avoid areas with many rooftop antennas.
        (3) Check the area of the location for local causes of reception interference, such as heavy motor vehicle traffic or high-tension power lines. For example, to detect such sources of interference, use a small AM transistor radio (with earphone) set to the highest frequency.
        (4) Check local available power for adequacy of voltage, current, and frequency requirements. Ensure power sources are sufficient to preclude fluctuations in other areas of the neighborhood when transmitting.
        (5) Check the immediate area around the location for metal obstructions that may cause a distortion in the radiation pattern of the transmitted signal.
        (6) Determine the bearing to the receiving station in relation to the layout of the location.
        (7) Check the location for adequate horizontal clearance in the direction of transmission in relation to the required transmission angle. The horizon, whether it is from terrain or buildings, should be no more than one-half the transmission angle.
        (8) Check the location for a means to conceal or disguise the antenna and transmission line from view and a means to support the antenna, such as exterior steps made of wood, guy wires for signs, rain gutters, or clotheslines.
    b. Determine the types of antennas that can be erected in the location using the available space, supports, and concealment areas.
    c. Construct and install selected antenna.
        (1) Construct the antenna using adequate insulators, when necessary, to prevent shorting of antenna. (Figure 3 shows expedient insulators.)
        (2) Conceal or disguise the antenna.
        (3) Conceal the transmission line from the location of the radio to the feed point on the antenna.
        (4) Check the concealment of the antenna and transmission line from different angles in the area. Ensure no evidence of the antenna, to include wire elements, transmission line, halyard, and insulators, is visible or unnatural to the location.
        (5) Connect the radio to the antenna, tune the radio, and transmit a signal, ensuring the antenna works properly.

Evaluation Preparation: Setup: Provide the Soldier with an HF radio set, unit SOI, installation sites (indoor and outdoor, urban and rural), designated frequency, pencil and paper, lensatic compass, expedient insulator material, electrical tape, knife, and pliers.

Performance Measures   GO   NO-GO

1. Constructed and installed an indoor antenna that is concealed or disguised and   ——   ——
   operational.
    a. Selected the most suitable indoor location at the installation site for the required transmission.
    b. Selected a suitable antenna for the indoor location and required transmission.
    c. Determined the length of antenna for the frequency given.
    d. Assembled the components.
    e. Connected the antenna to the radio, tuned it, and transmitted a signal.

STP 31-18E34-SM-TG

Performance Measures   GO   NO-GO

  2. Constructed and installed an outdoor antenna that was concealed or disguised   ___   ___
    and operational.
      a. Selected the most suitable outdoor location at the installation site for the
        required transmission.
      b. Selected a suitable antenna for the outdoor location and required
        transmission.
      c. Constructed the antenna, ensuring that adequate insulators were used to
        prevent shorting of the antenna.
      d. Concealed the antenna and transmission line.
      e. Checked the concealment of the antenna and transmission line.
      f. Connected the antenna to the radio, tuned it, and transmitted a signal.

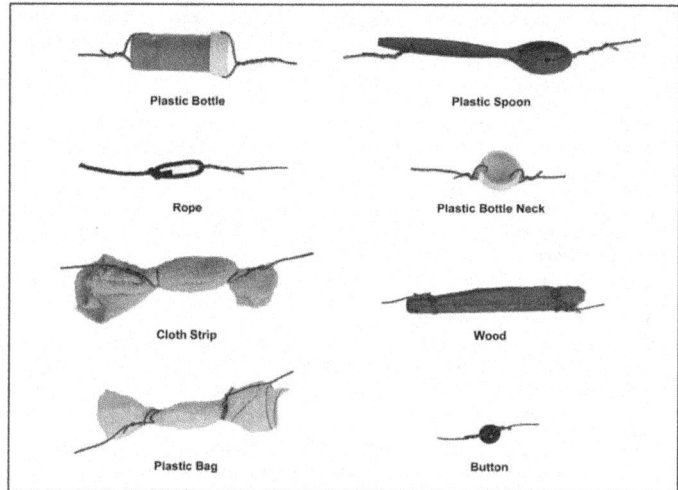

Figure 3. Expedient insulators

Evaluation Guidance: Score the Soldier GO if all performance measures are passed (P). Score the Soldier NO-GO if any performance measure is failed (F). If the Soldier fails any performance measure, show what was done wrong and how to do it correctly.

STP 31-18E34-SM-TG

Install Antenna Group OE-452/PRC
331-18E-3043

Conditions: Given at least two personnel, antenna group OE-452/PRC, communication range (distance), an installation site, an installed very-high-frequency (VHF) or high-frequency (HF) radio set, a compass, and Technical Manual (TM) 11-5985-391-12, *Operator's and Unit Maintenance Manual for Special Operations Radio Antenna Kit (SORAK): Antenna Group OE-452/PRC*. This task may be performed in a chemical, biological, radiological, or nuclear (CBRN) environment; therefore, some training should be done in mission-oriented protective posture (MOPP) 4.

Standards: Install antenna group OE-452/PRC into one of its 5 configurations in accordance with (IAW) TM 11-5895-391-12 within 30 minutes.

Performance Steps

1. Select an antenna configuration. (Refer to TM 11-5985-391-12, Table 2-1.)
2. Erect a 22-foot mast. (Refer to TM 11-5985-391-12.)
   a. Ensure the mast is vertical if deploying the 175-foot bent long-wire antenna.
   b. Ensure the mast is bent slightly beyond the vertical opposite the direction of pull of the first antenna wire deployed if deploying the 117-foot sloping dipole or 468-foot bent long-wire.
   c. Preload the top of the mast for an offset of 1 to 1.5 feet from the vertical if deploying a sloping-vee antenna.
3. Erect the antenna in the selected configuration.
   a. Deploy the 175-foot bent long-wire antenna. (Refer to TM 11-5985-391-12.)
      (1) Deploy the antenna wire.
      (2) Terminate the deployed antenna wire.
      (3) Connect the antenna to the radio (Figure 1).
   b. Deploy the 117-sloping dipole and 117-or 234-foot sloping-vee antennas. (Refer to TM 11-5985-391-12.)
      (1) Rig the antenna assembly at the 22-foot mast.
      (2) Deploy the antenna wire along the proper heading.
      (3) Erect the 6-foot mast.
      (4) Terminate the deployed antenna wire.
      (5) Repeat subparagraphs 3b(1) through 3b(4) to deploy the second leg of the antenna.
      (6) Connect the antenna to the radio.
   c. Deploy the 468-foot bent long-wire antenna. (Refer to TM 11-5985-391-12.)
      (1) Rig antenna assembly at the 22-foot mast.
      (2) Deploy the antenna wire along the proper heading.
      (3) Erect the 6-foot mast.
      (4) Terminate the deployed antenna wire.
      (5) Repeat subparagraphs 3c(1) through 3c(4) to deploy the second leg of the antenna.
      (6) Attach the balun, ground stake, and coax.
      (7) Connect the coax to the radio.

Evaluation Preparation: Setup: Provide the team of Soldiers with antenna group OE-452/PRC, communication range (distance), an installation site, an installed VHF or HF radio set, a compass, and TM 11-5985-391-12.

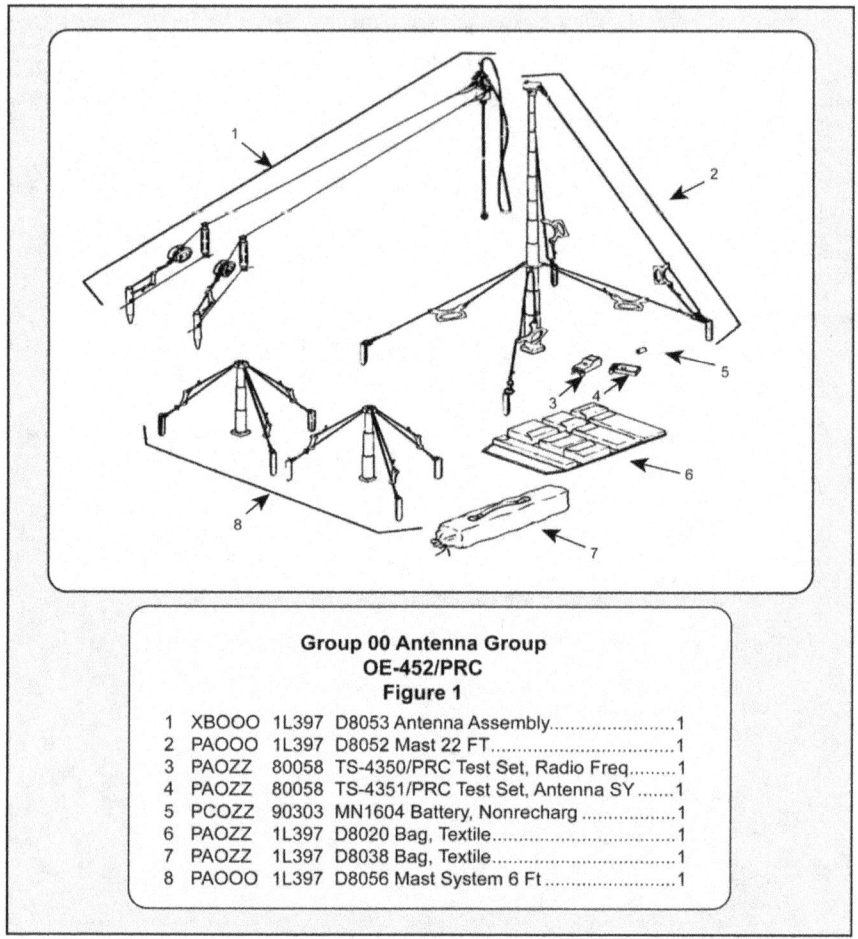

Group 00 Antenna Group
OE-452/PRC
Figure 1

```
1  XBOOO  1L397  D8053 Antenna Assembly........................1
2  PAOOO  1L397  D8052 Mast 22 FT....................................1
3  PAOZZ  80058  TS-4350/PRC Test Set, Radio Freq..........1
4  PAOZZ  80058  TS-4351/PRC Test Set, Antenna SY.......1
5  PCOZZ  90303  MN1604 Battery, Nonrecharg ..................1
6  PAOZZ  1L397  D8020 Bag, Textile....................................1
7  PAOZZ  1L397  D8038 Bag, Textile....................................1
8  PAOOO  1L397  D8056 Mast System 6 Ft .........................1
```

Figure 1. Antenna group OE-452

Performance Measures   <u>GO</u>   <u>NO-GO</u>

1. Selected an antenna configuration. (Refer to TM 11-5985-391-12, Table 2-1.)   ___   ___

2. Erected a 22-foot mast. (Refer to TM 11-5985-391-12.)   ___   ___
   a. Ensured the mast is vertical if deploying the 175-foot bent long-wire antenna.
   b. Ensured the mast is bent slightly beyond the vertical opposite the direction of pull of the first antenna wire deployed if deploying the 117-foot sloping dipole or 468-foot bent long-wire.

Performance Measures <u>GO</u>   <u>NO-GO</u>
    c. Preloaded the top of the mast for an offset of 1 to 1.5 feet from the vertical if deploying a sloping-vee antenna.

  3. Erected the antenna in the selected configuration.   ——— ———
    a. Deployed the 175-foot bent long-wire antenna. (Refer to TM 11-5985-391-12.)
    b. Deployed the 117-sloping dipole and 117- or 234-foot sloping-vee antennas. (Refer to TM 11-5985-391-12.)
    c. Deployed the 468-foot bent long-wire antenna. (Refer to TM 11-5985-391-12.)

Evaluation Guidance: Score the Soldier GO if all performance measures are passed (P). Score the Soldier NO-GO if any performance measure is failed (F). If the Soldier fails any performance measure, show what was done wrong and how to do it correctly.

References
  Required   **Related**
  TM 11-5985-391-12

STP 31-18E34-SM-TG

Construct Sloping-Vee Antenna
331-18E-3044

Conditions: Given 500 to 1000 feet of antenna wire, 10 feet of 600-ohm open-wire feed line, a balun, 100 feet of coaxial cable or 600-ohm open-wire feed line, two 300 to 400-ohm resistors, four insulators (or material to construct insulators), one 50 to 75-foot and two 15 to 30-foot antenna support masts (or site with existing supports, such as trees or structures), a measuring tape, 200 feet of guy-rope material, knife, pliers, suitable radio set, compass, writing material, three stakes, and current signal operating instructions (SOI).

Standards: Compute antenna length, assemble antenna, and connect antenna to radio set for operation within 30 minutes.

Performance Steps

1. Assemble the antenna (Figure 1).

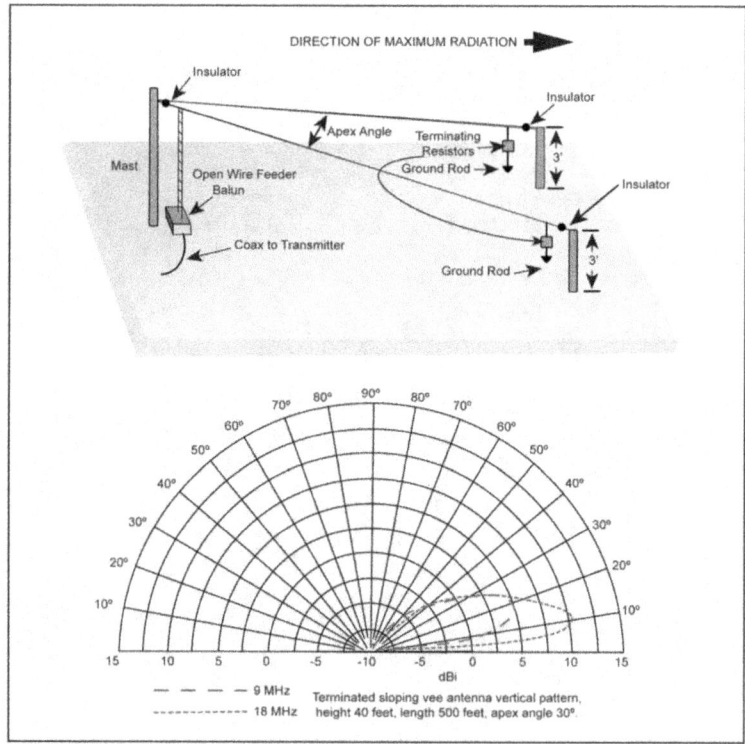

Figure 1. Sloping-vee antenna and Smith Chart

STP 31-18E34-SM-TG

Performance Steps
   Note: Characteristics of a sloping-vee antenna:
   - Frequency range: 3 to 30 megahertz (MHz)
   - Polarization: Horizontal
   - Power Cap: Dependent on terminating resistors
   - Radiation Pattern: 20 degrees either side of direction of radiation

Note: There are many variables to constructing any field-expedient antenna. The sloping-vee antenna should be at least one wavelength long and preferably several wavelengths long. A compromise tactical sloping-vee antenna can be constructed using two 500-foot legs. The length can be calculated by using the following formula:

$L = 492(N-.05)/\text{frequency (MHz)}$
$L = \text{length}$
$N = \text{number of half wavelengths}$

Note: This task uses the compromise tactical sloping-vee antenna for measurement purposes.

   a. Measure two lengths of antenna wire, each 502 feet long (add 12 inches to each end for insulators).
   b. Connect insulators to both ends of each antenna leg, ensuring the length between the insulators is 500 feet.
   c. Connect the insulator on one end of the legs to the insulator on the end of the remaining leg with the halyard material 4-to-6 inches apart.
   d. Connect the balun end assembly to antenna legs at the apex.
   e. Connect sufficient halyard material to the apex and terminating ends to raise the antenna to the proper height.

2. Erect the antenna.
   a. Determine azimuth to the distant station.
   b. Align terminating ends toward the direction of the receiving antenna.
   c. Erect three antenna masts or choose three trees at the site to be used as antenna masts.
   d. Elevate the apex of the antenna to a height of no more than 75 feet and no less than 49 feet.
   e. Adjust the distances between the legs (equal distance apart) to provide maximum radiation at the desired take-off angle.

Note: The following angles between the legs (apex angles) will give good results for the distances indicated:

   60 degrees: 700 to 1,000 miles
   45 degrees: 1,000 to 1,500 miles
   30 degrees: over 1,500 miles

   f. Elevate each of the terminating ends not more than 26 and not less than 15 feet.
   g. Attach a resistor to each end of the terminating ends of the antenna legs.
   h. Drive stakes into the ground and attach the halyard to the stakes.
3. Attach the antenna to the radio set.
   a. Attach the ends of the feed line to the antenna posts or the coaxial cable connector to the coaxial post on the radio set.

Evaluation Preparation: Setup: Provide the Soldier with 500 to1,000 feet of antenna wire, 10 feet of 600-ohm open-wire feed line, a balun, 100 feet of coaxial cable or 600-ohm open-wire feed line, two 300 to 400-ohm resistors, four insulators (or material to construct insulators), one 50 to 75-foot and two 15 to 30-foot antenna support masts (or site with existing supports, such as trees or structures), a measuring tape, 200 feet of guy-rope material, knife, pliers, suitable radio set, compass, writing material, three stakes, and current SOI.

STP 31-18E34-SM-TG

Performance Measures   <u>GO</u>   <u>NO-GO</u>

1. Assembled the antenna.
    a. Measured two lengths of antenna wire, each 502 feet long (added 12 inches to each end for insulators).
    b. Connected insulators to both ends of each antenna leg, ensuring the length between the insulators is 500 feet.
    c. Connected the insulator on one end of one leg to the insulator on the end of the other leg with the halyard material 4 to 6 inches apart.
    d. Connected the balun end assembly to antenna legs at the apex.
    e. Connected sufficient halyard material to the apex and terminating ends to raise the antenna to proper height.

2. Erected the antenna.
    a. Determined azimuth to the distant station.
    b. Aligned terminating ends toward the direction of the receiving antenna.
    c. Erected three antenna masts or chose three trees at the site to be used as antenna masts.
    d. Elevated the apex of the antenna to a height of no more than 75 feet and no less than 49 feet.
    e. Adjusted the distances between the legs (equal distance apart) to provide maximum radiation at the desired take-off angle.
    f. Elevated each of the terminating ends not more than 26 feet and not less than 15 feet.
    g. Attached a resistor to each end of the terminating ends of the antenna legs.
    h. Drove stakes into the ground and attached the halyard to the stakes.

3. Attached the antenna to the radio set.
    a. Attached the ends of the feed line to the antenna posts or the coaxial cable connector to the coaxial post on the radio set.

Evaluation Guidance: Score the Soldier GO if all performance measures are passed (P). Score the Soldier NO-GO if any performance measure is failed (F). If the Soldier fails any performance measure, show what was done wrong and how to do it correctly.

Note: Soldier needs to be trained in basic antenna theory, to include computing antenna lengths, and be able to determine direction with a compass.

References
    Required   **Related**
                                   FM 11-65

STP 31-18E34-SM-TG

Construct Expedient RC-292 Antenna
331-18E-3047

Conditions: Given 55 feet of W-1 antenna wire, 40 feet of coaxial cable RG-59, electrical tape, wire cutters, radio set, insulators (or material to construct field-expedient insulators), suspension line, a measuring device, signal operating instructions (SOI) with frequencies and call signs, ground plane spreaders (any rigid material), a suitable training site, and an assisting station.

Standards: Construct a field expedient RC-292 antenna and establish contact with an assisting station.

Performance Steps

1. Determine the length of the antenna for the frequency given. Length is determined by dividing the constant of 234 by the frequency in megahertz (MHz). For example: 234 divided by 55.15 MHz equals 4.24 feet (or 4 feet 3 inches).

2. Assemble the components.
   a. Using the calculations from step 1 and W-1 antenna wire provided, add 1 foot to the number obtained and cut 4 equal lengths of wire. Note: Use 6 inches of wire on each end of the component to connect the insulators.
   b. Select five issued insulators or construct five expedient insulators from materials provided.
      (1) Take each section of wire and measure 6 inches from each end.
      (2) Take one section of wire and connect an insulator to each end of this wire at the 6-inch mark. These are the ground plane elements.
      (3) Take the other three sections of wire and connect an insulator to one end of each of these sections at the 6-inch mark. These are the ground plane elements (Figure 1).
   c. Prepare the coaxial cable.
      (1) Take the coaxial cable and strip 10 inches of the black rubber coating from the cable, ensuring that the copper shielding underneath is not cut.
      (2) Push the copper shielding down to the black rubber coating. Near the rubber coating, spread the shielding apart to create an opening (hole).
      (3) Bend the coaxial cable at the opening just created and pull the plastic-coated center conductor through the hole.
      (4) Strip 5 inches of the plastic coating off the center conductor being careful not to nick or cut the thin center conductor.
   d. Construct the antenna.
      (1) Twist the three ground plane elements and the ground shield of the stripped coaxial cable together (6-inch twist).
      (2) Put the ground plane elements/ground shield twist through the bottom of the center insulator and wrap the excess wire of the twist around the ground plane elements close to the insulator.
      (3) Pass the plastic-coated center conductor through the top of the twist around the ground plane elements close to the insulator.
      (4) Using the ground plane spreaders and W-1 wire provided, connect the three sticks together to form a triangle. Note: Each leg of the triangle should be as long as the radiating element.
      (5) Connect each corner of the triangle to the insulator of a ground plane element with W-1 wire.

3. Erect the antenna.
   a. Using the suspension as a halyard, tie the halyard to the free end of the insulator of the radiating element.
   b. Connect a weight to the other end of the halyard and throw it over a tree limb.
   c. Pull the antenna up and secure the end of the halyard to the tree or other support.
      Note: Ensure the antenna center insulator is at least 30 feet above ground level and that no portion of the antenna touches vegetation.

8 February 2010

3-73

STP 31-18E34-SM-TG

Performance Steps

4. Attach the radio.
    a. Strip the other end of the coaxial cable as in paragraph 2c.
    b. Connect the center conductor to the antenna lead of the radio set.
    c. Connect the ground shield wire to the ground connector.
5. Establish contact with the assisting station.

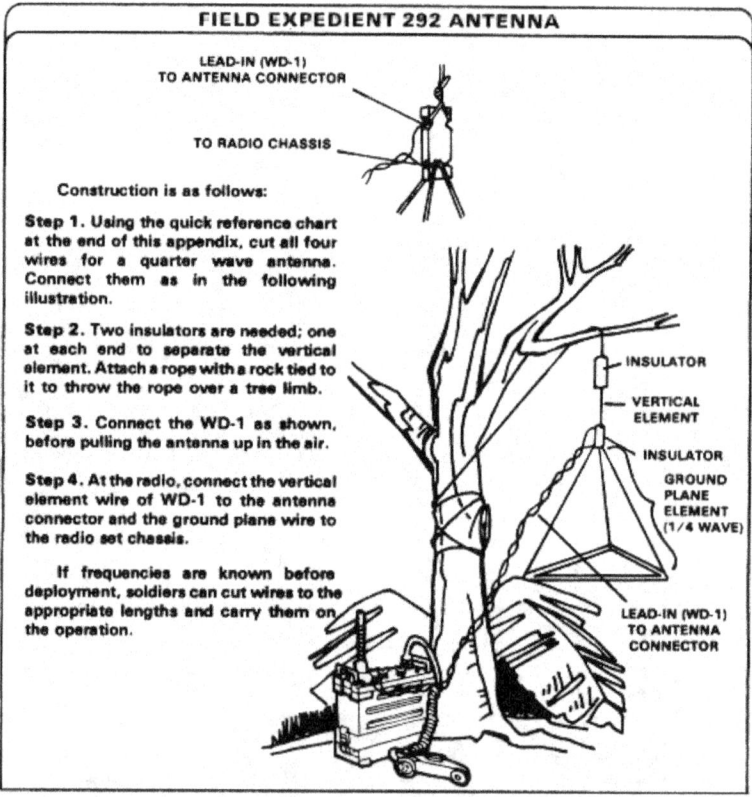

Figure 1. Expedient RC-292 antenna

Evaluation Preparation: Setup: Provide the Soldier with 55 feet of W-1 antenna wire, 40 feet of coaxial cable RG-59, electrical tape, wire cutters, radio set, insulators (or material to construct field-expedient insulators), suspension line, a measuring device, SOI with frequencies and call signs, ground plane spreaders (any rigid material), a suitable training site, and an assisting station.

STP 31-18E34-SM-TG

Performance Measures   <u>GO</u>   <u>NO-GO</u>

1. Determined the length of the antenna for the frequency given.   ——   ——
2. Assembled the components.   ——   ——
3. Erected the antenna.   ——   ——
4. Attached the radio.   ——   ——
5. Established contact with the assisting station.   ——   ——

Evaluation Guidance: Score the Soldier GO if all performance measures are passed (P). Score the Soldier NO-GO if any performance measure is failed (F). If the Soldier fails any performance measure, show what was done wrong and how to do it correctly.

References
    Required   **Related**
                                               UNIT SOI

STP 31-18E34-SM-TG

Employ Radio Set AN/PRC-137F
331-18E-3048

Conditions: Given a radio set AN/PRC-137F (complete), a terminating base station, and a communications requirement.

Standards: Install and operate the AN/PRC-137F radio.

Performance Steps

1. Check the radio for completeness and damage.

WARNING

The following protective measures should be observed when operating the radio in extreme weather conditions.

---

   a. Cold climates:
      (1) Use caution when handling and connecting cables to prevent kinks and unnecessary loops from causing permanent damage.
      (2) Make sure all connectors are free of frost, snow, and ice. Replace covers when not in use.
      (3) Never drag or place an open connector in the frost, snow, or ice.
   b. Hot climates:
      (1) Replace the connector covers as soon as the cable is disconnected.
      (2) Never place an open connector on the ground.
   c. In damp climates: Use a soft cloth to wipe all moisture and fungi from the equipment.

2. Install the battery.
   a. Verify that the mode switch is set to OFF position.
   b. Position the receiver/transmitter (R/T) unit, with the right side down, on a solid surface.
   c. Unfasten the two drawhook latches and remove the battery box.
   d. Inspect the battery box and the AA batteries, if installed, for signs of damage.
   e. Press the battery test switch to test the batteries.
   f. Replace the battery box and fasten the two drawhook latches.
   g. Connect the battery cable to the battery.

WARNING

A lithium-sulfur dioxide (Li-SO2) battery is used with the R/T. It contains pressurized sulfur dioxide (SO2) gas, which is toxic. The battery MUST NOT be abused in any manner that would cause it to rupture.

---

      (1) DO NOT heat, short-circuit, crush, puncture, mutilate, or disassemble the battery.
      (2) DO NOT use a battery showing signs of damage, such as bulging, swelling, disfigurement, a swollen plastic wrap, or a brown liquid in the plastic wrap.
      (3) DO NOT test a Li-SO2 battery for capacity.
      (4) DO NOT recharge a Li-SO2 battery. Lithium-based batteries may explode during recharging, causing severe personal injury and equipment damage.
      (5) DO NOT use water to extinguish Li-SO2 battery fires if a shock hazard exists due to

STP 31-18E34-SM-TG

Performance Steps

high-voltage electrical equipment in the vicinity (that is, greater than 30 volts, alternating current [VAC]; or volts, direct current [VDC]).
(6) If the battery becomes hot to the touch, if you hear a hissing sound (battery venting), or detect SO2 gas (a pungent, irritating odor), IMMEDIATELY turn off the equipment. Move the equipment to a well-ventilated area or leave the area.
(7) DO NOT use a halon-type fire extinguisher on a lithium battery fire.
(8) In the event of a fire near a lithium battery, rapid cooling of the battery is important. Use a carbon dioxide (CO2) extinguisher. Control of the equipment fire and battery cooling may prevent the battery from venting and potentially exposing lithium metal. If lithium becomes involved in the fire, the use of a graphite-based Class D fire extinguisher is recommended (for example, Lith-X or MET-L-X).
(9) DO NOT store lithium batteries with other hazardous materials. Keep them away from open flame or heat.

3. Install the antenna.
   a. Attach the ground or counterpoise wire to the black antenna binding post.
   b. Attach the antenna feed line to the green antenna binding post.
   c. Remaining accessories are installed during the operation of the radio set.

Note. The outstation (OS) radio is supplied with a Bayonet Neill Concelman (BNC) adapter for interfacing to standard antenna systems (for example, AS-2259, special operations radio antenna kit [SORAK], dipoles, and whips). The coupler eliminates the need for the antenna to present 50-ohm impedance to the OS.

4. Initialize the AN/PRC-137F.
   a. Inventory and inspect equipment.
   b. Rotate the mode selector switch to the OFF position.
   c. Connect the battery cable to the external battery.
   d. Rotate the mode selector switch to a channel position.
   e. Randomize the radio.
      (1) Connect a handset to the red dot connector.
      (2) Connect the handset to the red dot connector.
      (3) Observe that the light emitting diode (LED) momentarily stops flashing.
      (4) Disconnect the handset.
   f. Click on the digital message device (DMD) icon on the desktop screen to open the DMD program.
   g. From the main menu select FILE and INITIALIZE.
   h. Set/verify the correct date and time (ZULU) and cryptographic (crypto) month and year.
   i. Load the crypto.
   j. Upload the parameters.
   k. Set the advanced parameters.
      (1) Verify/set ALE group.
      (2) Enter text alias.
      (3) Verify adaptive power control and standby mode settings.
      (4) Set wake interval.
   l. Enter correct base identification (ID).
   m. Rotate mode selector switch to the NETWORK position. Verify the radio is tuning and wait for a net entry.
   n. Verify net entry.
      (1) Rotate the mode selector switch to a channel position.
      (2) Observe for appearance of the base identification in the contacted bases window.

Note. Network entry may also be verified by clicking on the base icon button and looking for the checked contacted box.

STP 31-18E34-SM-TG

Performance Steps
Note. The time of day and date should be checked against the AN/TRQ-43G for accuracy prior to deployment. The OS date and time must be within two minutes of the base station date and time in order to be operational. It is recommended the OS be tested over the air with the base station prior to deployment.

    5. Operate the AN/PRC-137F.
        a. Turn on the DMD. Rotate the selector switch to a channel position.
        b. Click on the NEW message button and type in the message.
        c. After the complete message text is typed, click the SEND button.
        d. Rotate the mode selector switch to the NETWORK position.

    6. Zeroize the AN/PRC-137F.
        a. Rotate the mode selector switch to a channel position.
        b. Click on the ZEROIZE button.

        c. Select the desired zero procedure.
            (1) Soft: Erases black keys and messages in the radio.
            (2) Hard: Erases crypto and messages in the radio.
            (3) System: Erases all crypto and messages in the radio and messages from the DMD mail folders.
        d. Follow the on-screen instructions.

Evaluation Preparation: Setup: Provide the Soldier with a radio set AN/PRC-137F (complete), a terminating base station, and a communications requirement.

Performance Measures	GO	NO-GO
1. Checked the radio for completeness and damage.	——	——
2. Installed the battery.	——	——
3. Installed the antenna.	——	——
4. Initialized the AN/PRC-137F.	——	——
5. Operated the AN/PRC-137F.	——	——
6. Zeroized the AN/PRC-137F.	——	——

Evaluation Guidance: Score the Soldier GO if all performance measures are passed (P). Score the Soldier NO-GO if any performance measure is failed (F). If the Soldier fails any performance measure, show what was done wrong and how to do it correctly.

STP 31-18E34-SM-TG

Employ Radio Set AN/PSC-5D
331-18E-3052

Conditions: Given an AN/PSC-5D multiband multimission radio (MBMMR), a communications requirement, operating frequency, distant station, and Technical Manual (TM) 11-5820-1130-12&P (PSC-5D).

Standards: Install, operate, and maintain radio set in accordance with (IAW) TM 11-5820-1130-12&P, and establish communications with distant station at scheduled contact time or within 30 minutes.

Performance Steps
Warnings: Two lithium-sulfur dioxide (Li-SO2) batteries are used with the AN/PSC-5D. They contain pressurized sulfur dioxide (SO2) gas, which is toxic. The batteries MUST NOT be abused in any manner that would cause them to rupture.
- DO NOT heat, short circuit, crush, puncture, mutilate, or disassemble the batteries.
- DO NOT use a battery showing signs of damage, such as bulging, swelling, disfigurement, a swollen plastic wrap, or a brown liquid in the plastic wrap.
- DO NOT test Li-SO2 batteries for capacity.
- DO NOT recharge Li-SO2 batteries. Lithium-based batteries may explode during recharging, causing severe personal injury and equipment damage.
- DO NOT use water to extinguish Li-SO2 battery fires if a shock hazard exists due to high voltage electrical equipment in the vicinity (that is, greater than 30 volts, alternating current [VAC], or direct current [VDC]).
- If the battery compartment becomes hot to the touch, if you hear a hissing sound (battery venting), or detect SO2 gas (a pungent, irritating odor), IMMEDIATELY turn off the equipment. Move the equipment to a well-ventilated area or leave the area.
- DO NOT use a halon-type fire extinguisher on a lithium battery fire.
- In the event of a fire near a lithium battery, rapid cooling of the battery is important. Use a carbon dioxide (CO2) extinguisher. Control of the equipment fire, and cooling, may prevent the battery from venting and potentially exposing lithium metal. If lithium becomes involved in the fire, the use of a graphite-based Class D fire extinguisher is recommended; for example, Lith-X, MET-L-X.
- DO NOT store lithium batteries with other hazardous materials. Keep them away from open flame or heat.

1. Conduct an inventory of the MBMMR AN/PSC-5D (manpack configuration) for completeness and serviceability (Figure 1).
    a. Manpack radio equipment.
        (1) Receiver/transmitter (R/T), RT-1672D(C)/U.
        (2) Battery BB-390A/U.
        (3) Battery box assembly.
        (4) Cloning cable assembly.
        (5) Remote data adapter cable assembly.
        (6) Remote data user cable assembly.
        (7) Global positioning system (GPS) time of day (TOD) cable assembly.
        (8) Backpack accessory bag.
        (9) Portable remote-control device (PRCD).
        (10) Handset H-250/U.
        (11) Flexible very-high frequency (VHF) antenna (30 to 88 megahertz [MHz]).
        (12) Flexible VHF/ultra-high frequency (UHF) antenna (30 to 512 MHz).
    b. AN/PSC-5D serviceability checks.
        (1) Check control knobs to ensure they are tight.
        (2) Check the displays for cracks and breakage.
        (3) Check connectors for bent or missing pins, distorted or broken shells.

STP 31-18E34-SM-TG

Performance Steps

    (4) Check the handset for broken or missing parts. Operate the push-to-talk (PTT) switch and inspect the cable connector.
    (5) Manually check the battery box latches and the pressure relief valve for proper operation.
    (6) Manually check the R/T latch catch for proper operations.
    (7) Check the rubber gasket on rear of R/T; ensure it is pliable and not damaged.
    (8) Check the interface cables for signs of damage.
    (9) Check the interface cable connections for bent or missing pins and distorted or broken shells.
    (10) Check the line of sight (LOS) antenna(s) connector, shell, and body for signs of damage.
    (11) Check the satellite antenna assembly, including radio frequency (RF) connector shell, and body for signs of damage.

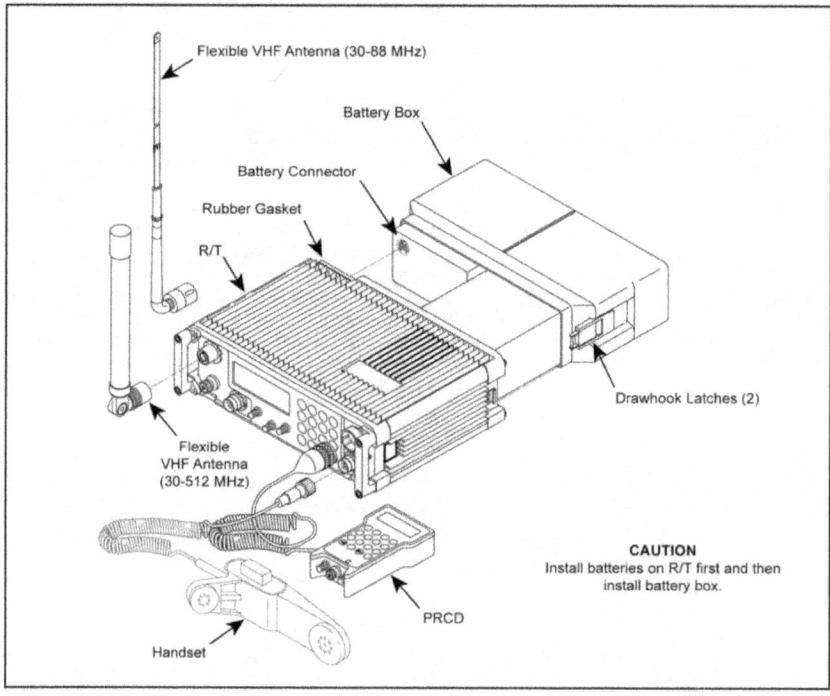

Figure 1. AN PSC 5D breakout

Performance Steps

2. Install battery (Figure 2).
    a. Verify the radio set mode switch is set to the OFF position.
    b. Unfasten the drawhook latches and remove the battery box.
    c. Inspect the battery box and the batteries, if installed, for signs of damage.
    d. Disconnect and replace the batteries one at a time.
    e. Replace the battery box.

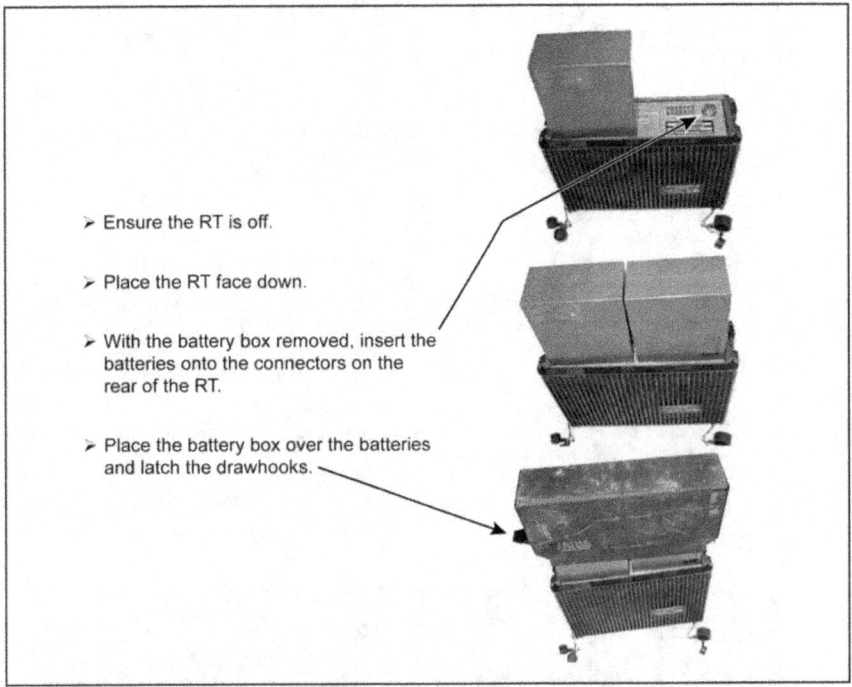

- Ensure the RT is off.
- Place the RT face down.
- With the battery box removed, insert the batteries onto the connectors on the rear of the RT.
- Place the battery box over the batteries and latch the drawhooks.

Figure 2. Battery Installation

3. Install accessories (Figure 3).
    a. Install the radio in the user-supplied carrying harness.
    b. Connect the LOS antenna to the antenna connector on the radio. If using the satellite antenna, connect the antenna cable to antenna connector.
    c. If using a personal remote controller device, connect the PRCD cable to the auxiliary (AUX) connector on the radio or to the remote data adapter cable, if used.
    d. If using the smart remote control—
        (1) Connect the P1 of the remote data user cable (part number [PN] 432204-801) to the remote/data adapter (PN 423155-1), if in use, or directly to the AUX connector on the radio.

STP 31-18E34-SM-TG

Performance Steps

(2) Connect the P3 of the remote data user cable to the serial port on the personal computer (PC).

Note. The remote data user cable can also be connected to the radio through the remote data adapter.

4. Install the global positioning receiver. Connect the GPS receiver directly to the AUX connector on the radio or to the remote data adapter cable, if used.

5. Connect external devices. If using external input/output (I/O) message devices (for example, KL-43C/KL-43F, AN/PSC-2, OA-8990), connect the device to the remote data adapter cable or directly to the AUX connector on the radio.

6. Conduct manpack power-up procedures (Figure 4).
   a. Ensure the batteries are installed.
   b. If operating from an R/T, set the mode switch to plaintext (PT).
   c. If operating from the PRCD, perform the following steps:
      (1) Set the mode switch on the R/T to remote (RMT) position.
      (2) Set the mode switch on the PRCD to the PT position. After the opening screen, the power-up messages on the PRCD are the same as the R/T's. The R/T displays, "THIS TERMINAL IS UNDER REMOTE CONTROL."
      (3) To adjust the PRCD backlight intensity or contrast, rotate the volume control fully counterclockwise to the detent position. Then use the arrow keys to adjust the backlight intensity or the NEXT/PREV key to adjust the contrast (press and release keys continuously to adjust). Rotate the volume control clockwise out of the detent position when finished.
   d. If operating from smart remote control (via personal computer [PC]) perform the following steps:
      (1) Set the mode switch on R/T to RMT position.
      (2) Boot up the PC and install the compact disk, read-only memory (CD-ROM), containing the smart remote control software.
      (3) Follow the prompts on the PC to operate the radio.

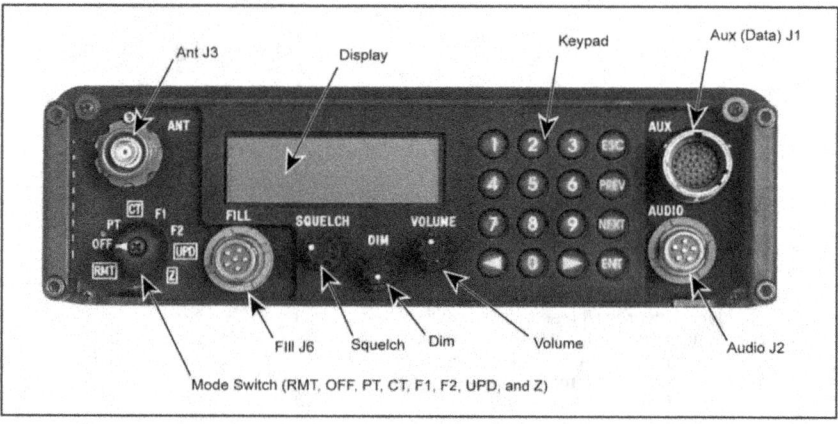

Figure 3. Front panel

3-82　　　　　　　　　　　　　　　　　　　　　　　　　　　　　　　　　　　　8 February 2010

STP 31-18E34-SM-TG

Performance Steps

7. Set the backlight timer.
    a. Press the ESC key as necessary until the main menu is displayed.
    b. From the main menu, press the 4 key to enter the configuration menu.
    c. From the configuration menu, press the 1 key to enter the terminal data menu.
    d. Press the NEXT/PREV key to move the cursor to the backlight timer field.
    e. Press the keypad number keys to select the backlight shut-off time in delay seconds (00-60).
    f. Press the ENTER key.
    g. Press the ESC key, as required, to return to the main menu.
8. Conduct manpack power-down procedures (Figure 5).
    a. Zero variables, if required, by setting mode switch to Z position (pull out knob to rotate). The display will show "ZEROIZED" when Z is selected.
    b. If operating from the PRCD, set the mode switch on the PRCD to the OFF position.
    c. If operating from the smart remote control, power down the PC.
    d. Set the mode switch on the R/T to the OFF position.

Evaluation Preparation: Setup: Provide the Soldier with an AN/PSC-5D MBMMR, a communications requirement, operating frequency, distant station, and TM 11-5820-1130-12&P.

Performance Measures	GO	NO-GO
1. Inventoried the AN/PSC-5D (manpack), for completeness and serviceability.	——	——
2. Installed the batteries.	——	——
3. Installed the necessary equipment for transmission; that is, handset, flexible VHF or satellite antenna, PRCD, smart remote control, and GPS receiver.	——	——
4. Conducted the power-up procedures.	——	——
5. Set the backlight time and radio settings for transmission.	——	——
6. Transmitted a message to the distant station.	——	——
7. Conducted proper shutdown procedures.	——	——
8. Broke down the radio set and equipment.	——	——

Evaluation Guidance: Score the Soldier GO if all performance measures are passed (P). Score the Soldier NO-GO if any performance measure is failed (F). If the Soldier fails any performance measure, show what was done wrong and how to do it correctly.

References
    Required   **Related**
    PN: MX-64-356
    TM 11-5820-1130-12&P

STP 31-18E34-SM-TG

Employ Radio Set AN/PRC-148
331-18E-3056

Conditions: Given a multiband intrateam radio (MBITR) with the Thales Operation Manual (TOM) or Quick Reference Guide (QRG).

Standards: Install and place the MBITR into operation in accordance with (IAW) the TOM. Conduct communications check with distant station IAW the contact time listed in the unit signal operating instructions (SOI) or within 5 minutes.

Performance Steps
Safety Summary. The following are general safety precautions not related to any specific procedure. These safety summaries are recommended precautions all personnel must understand and apply during any given phase of operation and maintenance.

CAUTION

Hazards of electromagnetic radiation to ordnance (HERO). DO NOT operate the radio within 10 feet (3 meters) of any type of fuzed ordnance. Operating the radio near ordnance MAY induce or otherwise couple currents or voltages of magnitudes large enough to initiate electro-explosive devices or other sensitive explosive components of weapon systems, ordnance, or explosive devices CAUTION: Lithium-ion batteries. Lithium-ion batteries have a very high energy density. Use caution when handling and testing. DO NOT short-circuit, overcharge, crush, or mutilate the battery. DO NOT penetrate the battery with a nail. DO NOT apply reverse polarity to the battery. DO NOT expose it to high temperature or disassemble it. High case temperature resulting from abuse of the cell could cause physical injury.

---

1. Inventory AN/PRC-148 and accessories.
   a. MBITR receiver transmitter (R/T) unit.
   b. Rechargeable lithium-ion batteries (2).
   c. Battery holder for nonrechargeable batteries (2).
   d. Receiver/transmitter (R/T) holster.
   e. Accessory carrying bag.
   f. Antenna complement 30 to 90 megahertz (MHz) through 30 to 512 MHz.

2. Place the AN/PRC-148 into operation.
   a. Install the battery.
   b. Install the correct antenna.
   c. Turn on the radio.
   d. Select the proper channel.
   e. Transmit a message to a distant station.

Evaluation Preparation: Setup: Provide the Soldier with an MBITR and the TOM or the QRG.

Performance Measures	GO	NO-GO

1. Inventoried AN/PRC-148 and accessories.
   a. MBITR R/T unit (Figure 1).
   b. Rechargeable lithium-ion batteries (2) (Figure 2).
   c. Battery holder for nonrechargeable batteries (2) (Figure 3).
	——	——

STP 31-18E34-SM-TG

Performance Measures   **GO**   **NO-GO**
    d. R/T holster (Figure 4).
    e. Accessory carrying bag (Figure 5).
    f. Antenna complement 30 to 90 MHz (Figure 6) through 90 to 512 MHz (Figure 7).

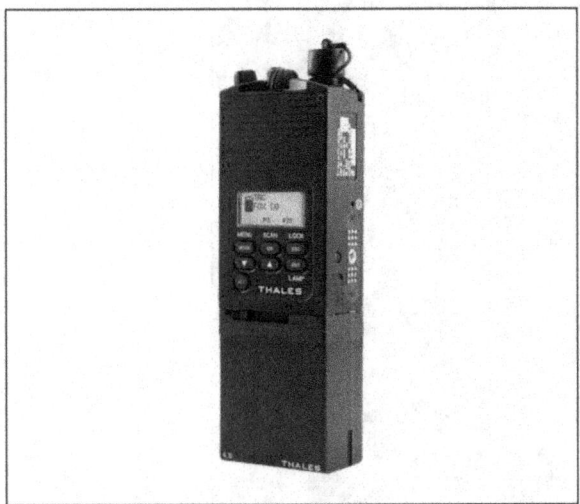

Figure 1. AN/PRC-148

Figure 2. Lithium-ion battery

Performance Measures GO   NO-GO

Figure 3. Battery holder

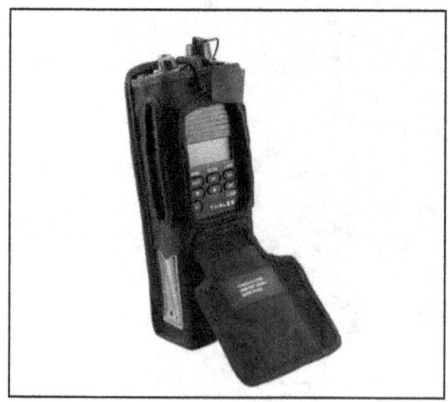

Figure 4. R/T holster

STP 31-18E34-SM-TG

Performance Measures GO NO-GO

Figure 5. Accessory carrying bag

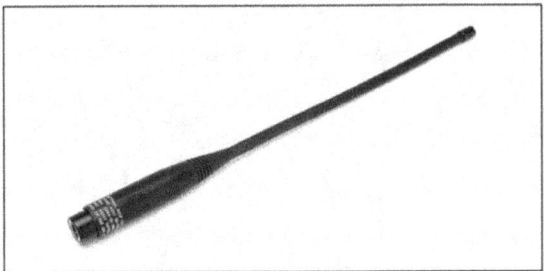

Figure 6. Antenna, 30 to 90 MHz

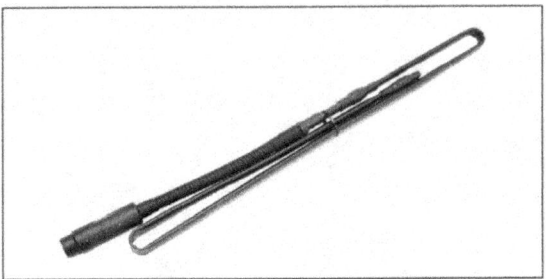

Figure 7. Antenna, 90 to 512 MHz

STP 31-18E34-SM-TG

Performance Measures   <u>GO</u>   <u>NO-GO</u>

    2. Placed the AN/PRC-148 into operation.
        a. Installed battery.
        b. Installed correct antenna.
        c. Turned on radio.
        d. Selected proper channel.
        e. Transmitted message to distant station.

Evaluation Guidance: Score the Soldier GO if all performance measures are passed (P). Score the Soldier NO-GO if any performance measure is failed (F). If the Soldier fails any performance measure, show what was done wrong and how to do it correctly.

STP 31-18E34-SM-TG

Employ Iridium Emergency Communications
331-18E-3070

Conditions: In a training, field, or combat environment, given a Motorola Iridium portable satellite phone, subscriber identity module (SIM) card, or a prepaid Iridium satellite phone, battery, antenna, and an Iridium quick reference sheet.

Standards: Employ Iridium phone and provide emergency satellite telephone communications to a distant station.

Performance Steps

1. Insert SIM card into the Iridium.
2. Remove the cover from the battery (Figure 1).

Figure 1. Battery cover removed

3. Install the battery onto the back of the Iridium.
4. Attach the battery compartment cover.
5. Charge the battery.
6. Attach the antenna to the antenna connection port.
7. Turn the phone ON.
8. Select preinserted phone number or manually dial distant station number (Figure 2).

STP 31-18E34-SM-TG

Performance Steps

Figure 2. Detailed keypad

Evaluation Preparation: Setup: Provide the Soldier with a Motorola Iridium portable satellite phone, SIM card, or a prepaid Iridium satellite phone, battery, antenna, and an Iridium quick reference sheet.

Performance Measures	GO	NO-GO
1. Inserted SIM card into the Iridium.	——	——
2. Removed cover from the battery.	——	——
3. Installed battery onto the back of the Iridium.	——	——
4. Attached the battery compartment cover.	——	——
5. Charged the battery.	——	——
6. Attached the antenna to the antenna connection port.	——	——
7. Turned the phone ON.	——	——
8. Selected preinserted phone number or manually dialed the distant station number.	——	——

Evaluation Guidance: Score the Soldier GO if all performance measures are passed. Score the Soldier NO-GO if any performance measure is failed. If the Soldier fails any performance measure, show what was done wrong and how to do it correctly.

References
    Required   Related
    9505A PORTABLE PHONE

STP 31-18E34-SM-TG

Construct a Slant-Wire Antenna
331-18E-3073

Conditions: Given 250 feet of antenna wire, issue-type insulators (or field-expedient material for constructing insulators), 150 feet of suspension line (or similar type halyard), knife, pliers, measuring device, compass, azimuth to orient antenna, a radio set, operating frequency, and a suitable installation site.

Standards: Cut antenna to operating frequency; assemble, construct, and connect antenna to the radio set. Task is complete when the radio set indicator indicates antenna is operational.

Performance Steps

1. Determine length of antenna for frequency given.
   a. Divide 234 by the frequency in megahertz (MHz) to determine the length of the radiating element.
   b. Determine length of the counterpoise to be used.

Note: Counterpoise must be at least as long or longer than the radiating element.

2. Assemble components.
   a. Using the calculations from above and the antenna wire provided, cut the correct lengths of wire.

Note: Soldier should add 6 inches to each end of wire to connect the insulators.

   b. Select issued insulators or construct insulators using field expedient techniques.
   c. Connect insulators to the wire.

3. Construct antenna (Figure 1).
   a. Antenna with counterpoise.
      (1) Set up radiating element on the azimuth to the base station, maintaining a 30 to 60-degree angle from horizontal.
      (2) Set up counterpoise on the azimuth to the base station.
   b. Antenna with ground stake.
      (1) Set up radiating element on the azimuth to the base station, maintaining a 30 to 60-degree angle from horizontal.
      (2) Drive stake into ground.

4. Connect antenna to radio.
   a. Connect radiating element to the radio binding post or screws under the antenna base.
   b. Connect the counterpoise to the radio binding post of to the radio chassis.

Evaluation Preparation: Setup: Provide the Soldier with 250 feet of antenna wire, issue-type insulators (or field-expedient material for constructing insulators), 150 feet of suspension line (or similar type halyard), knife, pliers, measuring device, compass, azimuth to orient antenna, a radio set, operating frequency, and a suitable installation site.

Performance Measures	GO	NO-GO
1. Determined the length of antenna for the frequency given.	——	——
2. Assembled the components.	——	——
3. Constructed the antenna.	——	——
4. Connected the antenna to the radio.	——	——

STP 31-18E34-SM-TG

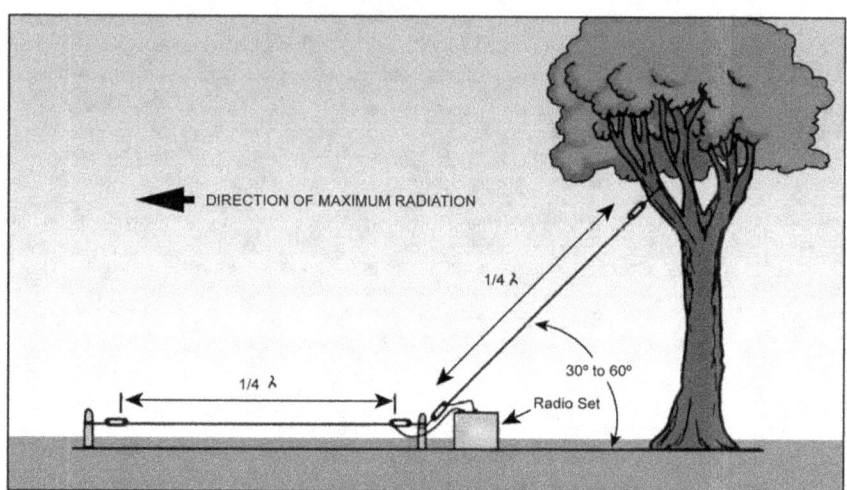

Figure 1. Slant-wire antenna

Evaluation Guidance: Score the Soldier GO if all performance measures are passed (P). Score the Soldier NO-GO if any performance measure is failed (F). If the Soldier fails any performance measure, show what was done wrong and how to do it correctly.

STP 31-18E34-SM-TG

Subject Area 4: Communications Planning

Plan Frequency Modulated Voice and Data Communications Net
113-611-6002

Conditions: Given Technical Manual (TM) 11-5820-890-20-1, *Unit Maintenance Manual for Ground ICOM Radio Sets AN/PRC-119A*; TM 11-5820-890-20-2, *Unit Maintenance Manual for Ground ICOM Radio Sets*; list of equipment (single-channel ground and airborne radio system [SINCGARS] radio set, AN/CYZ-10, Enhanced Position Location and Reporting System [EPLRS] radio set AN/VSQ-2(V)2), and unit's signal operating instructions (SOI), standing operating procedure (SOP), and operation plan (OPLAN).

Standards: Incorporated the voice and data communications net plan into the OPORD, and the commander approved the plan.

Performance Steps

1. Identify mission requirements of the frequency-modulated (FM) voice and data communications net.
    a. Review the OPLAN.
    b. Review the assets.

2. Formulate signal estimate.
    a. Develop the transmission plan diagram.
    b. Identify the sustainment requirements.

3. Plan for the following considerations:
    a. Terrain.
    b. Weather.
    c. Type of unit.
    d. Number of units.
    e. Availability of communications security (COMSEC) devices.
    f. Restricted frequencies.
    g. Antenna requirements.
    h. Power requirements.
    i. Electronic warfare (EW) threats.

4. Submit the plan for approval.
    a. Brigade or battalion signal officer.
    b. Training operations S3.

5. Incorporate the approved plan into the OPORD.

Performance Measures   <u>GO</u>   <u>NO-GO</u>

1. Identified mission requirements of the FM voice and data communications net.   ——   ——
    a. Reviewed the OPLAN.
    b. Reviewed the assets.

2. Formulated signal estimate.   ——   ——
    a. Developed the transmission plan diagram.
    b. Identified the sustainment requirements.

3. Planned for the following considerations: ——   ——
    a. Terrain.
    b. Weather.
    c. Type of units.
    d. Number of units.

8 February 2010   3-93

STP 31-18E34-SM-TG

Performance Measures   <u>GO</u>   <u>NO-GO</u>
    e. Availability of COMSEC devices.
    f. Restricted frequencies.
    g. Antenna requirements.
    h. Power requirements.
    i. EW threats.

4. Submitted the plan for approval.   ____   ____
    a. Brigade or battalion signal officer.
    b. Training operations S3.

5. Incorporated the approved plans into the OPORD.   ____   ____

Evaluation Guidance: Score the Soldier GO if all performance measures are passed (P). Score the Soldier NO-GO if any performance measure is failed (F). If the Soldier fails any performance measure, show what was done wrong and how to do it correctly.

References
    Required   **Related**
    TM 11-5820-890-20-1
    TM 11-5820-890-20-2
    UNIT OPLAN
    UNIT SOI
    UNIT SOP

STP 31-18E34-SM-TG

Plan for the Employment of Tactical Radio Systems
331-18E-3081

Conditions: Given a tactical situation, an operation order (OPORD) or operation plan (OPLAN), a unit standing operating procedure (SOP), and Field Manual (FM) 24-1, *Signal Support in the Airland Battle*.

Standards: Plan for the employment of tactical radio systems for the mission which meet communication requirements in accordance with (IAW) the OPORD or OPLAN.

Performance Steps

1. Review the OPORD or OPLAN.
    a. Determine the types of missions.
    b. Determine the size of unit to be supported.
2. Determine the radio systems to be installed.
3. Determine the availability of personnel and equipment needed to install the system(s).
4. Prepare diagrams, overlays, or schematics required for the system(s).
5. Forward the completed diagrams, overlays, or schematics for the system(s) to superior for approval.

Evaluation Preparation: Setup: Provide the Soldier with an OPORD or OPLAN, SOP, signal operating instructions (SOI), and applicable FMs.

Performance Measures	GO	NO-GO

1. Reviewed the OPORD or OPLAN.
   a. Determined the types of missions.
   b. Determined the size of unit to be supported.
	——	——

2. Determined the radio systems to be installed.	——	——
3. Determined the availability of personnel and equipment needed to install the system(s).	——	——
4. Prepared diagrams, overlays, or schematics required for the system(s).	——	——
5. Forwarded the completed diagrams, overlays, or schematics for the system(s) to superior for approval.	——	——

Evaluation Guidance: Score the Soldier GO if all performance measures are passed (P). Score the Soldier NO-GO if any performance measure is failed (F). If the Soldier fails any performance measure, show what was done wrong and how to do it correctly.

References
  Required   Related
  FM 24-1

STP 31-18E34-SM-TG

Install Mission Planning Kit
331-18E-3083

Conditions: Given an operational mission planning kit (MPK) (complete), software, network access, and applicable reference materials in a mission planning environment.

Standards: The MPK is installed and is functional for all users in accordance with (IAW) unit requirements.

Performance Steps

1. Inventory MPK.

2. Design local area network (LAN).

3. Implement LAN design.
    a. Connect computers to the switch with category V (CAT-V) cable.
    b. Configure computers.
        (1) Assign computer names.
        (2) Assign internet protocol (IP) addresses.
        (3) Synchronize system component time within 1 minute.

4. Share system resources.
    a. Share folders and drives.
    b. Share printer.
    c. Map network drives.
    d. Map external drive(s).

5. Configure peripherals.
    a. Configure scanner.
    b. Configure projector.

Evaluation Preparation: Provide the Soldier with all the material and equipment listed in the condition statement.

Performance Measures	GO	NO-GO
1. Inventoried MPK.	——	——
2. Designed a LAN.	——	——

3. Implemented LAN design.
   a. Connected computers to the switch with CAT-V cable.
   b. Configured computers.
      (1) Assigned computer names.
      (2) Assigned IP addresses.
      (3) Synchronized system component time within +/- 1 minute.
	——	——

4. Shared system resources.
   a. Shared folders and drives.
   b. Shared printer.
   c. Mapped network drives.
   d. Mapped external hard drive(s).
	——	——

5. Configured peripherals.
   a. Configured scanner.
   b. Configured projector.
	——	——

STP 31-18E34-SM-TG

Evaluation Guidance: Score the Soldier GO if all performance measures are passed (P). Score the Soldier NO-GO if any performance measure is failed (F). If the Soldier fails any performance measure, show what was done wrong and how to do it correctly.

STP 31-18E34-SM-TG

Skill Level 4

Subject Area 2: Communication Procedures

Establish a Special Operations Task Force Signal Center
331-18E-4003

Conditions: Given organic signal element with personnel and equipment, a unit standing operating procedure (SOP), unit signal operating instructions (SOI), radio net diagrams, telephone traffic diagrams, patchwork sheet, wiring plan, an operation order (OPORD), applicable equipment technical manuals (TMs), and FM 20-3, *Camouflage, Concealment, and Decoys.*

Standards: Camouflage the signal center, establish the site defense, and ensure the communications systems are ready to pass traffic within 3 hours.

Performance Steps

1. Review mission, SOP, communication diagrams, site defense plan, and SOI to determine the specific requirements for the installation signal center.

2. Implement the site defense plan.
    a. Position security outposts around site to provide early warning of an enemy approach.
    b. Establish entrance and exit points and lanes for traffic flow within the site.
    c. Coordinate with supported unit for specific defense responsibility and perimeter sector responsibility, as required.
    d. Coordinate with engineer element for assistance in establishing field fortifications for communications assemblages, as required.
    e. Direct location and construction of individual and crew-served fighting positions.
    f. Direct installation of artificial obstacles, as required.
    g. Direct installation of chemical alarm system, as required.
    h. Identify and locate focal point for command and control of site defense.
    i. Plan for orderly withdrawal.
    j. Prepare defense overlay, if required.

3. Direct installation of secure radioteletype (RATT) station. (Refer to applicable TMs.)
    a. Review the mission, SOP, SOI, and radio net diagram to determine specific requirements for installation and operation of a RATT terminal.
    b. Direct installation of RATT sets, to include organic power generators, if required.
    c. Direct antenna installation.
    d. Direct preoperational checks and adjustments on equipment, as required.
    e. Direct establishment of physical security.

4. Direct installation of high-frequency (HF) outstation radio net. (Refer to applicable TMs.)
    a. Review the mission, SOP, SOI, and radio net diagram to determine specific requirements for installation and operation of the base-station radio system.
    b. Direct installation of base-station radio station system, to include organic power generator, as required.
    c. Direct antenna installation.
    d. Direct preoperational checks and adjustments on equipment, as required.
    e. Direct establishment of physical security.

5. Direct installation of telephone central office. (Refer to equipment TM.)
    a. Review the mission, SOP, telephone traffic diagram, patchwork sheets, and SOI to determine specific requirements for installation and operation of manual telephone central office.
    b. Direct the installation of central manual telephone office.
    c. Direct the preparation of main distribution frame line and trunk cable records, as required.
    d. Direct the testing of line and trunk circuits.

Performance Steps
- e. Review the mission, SOP, telephone traffic diagram, patchwork sheet, program worksheet, and SOI to determine specific requirements for installation and operation of an automatic telephone central office.
- f. Direct establishment of central automatic telephone office.
- g. Review program worksheet (wiring plan), as required.
- h. Direct preliminary adjustments of switchboard.

6. Direct installation of internal and external cable or field.

7. Direct installation of a telecommunications center.
    - a. Review the mission, SOP, RATT traffic diagram, and initial messenger schedules to determine specific requirements for installation telecommunications center.
    - b. Direct installation of telegraph terminal and organic power generator.
    - c. Direct installation of message center equipment.
    - d. Establish messenger service, if required.
    - e. Direct the establishment of physical security.

8. Direct implementation of cover and concealment for personnel and equipment, as required. (Refer to FM 20-3.)

9. Coordinate with support personnel for petroleum, oils, lubricants (POL), mess facilities, administrative needs of assigned personnel, maintenance, and field sanitation, as required.

10. Coordinate with the communications-electronics (CE) officer/noncommissioned officer for support of nonorganic communications requirements.

Evaluation Preparation: Setup: Provide the Soldier with organic signal element with personnel and equipment, a unit SOP, unit SOI, radio net diagrams, telephone traffic diagrams, patchwork sheet, wiring plan, an OPORD, applicable equipment TMs, and FM 20-3.

Performance Measures	GO	NO-GO
1. Reviewed mission, SOP, communication diagrams, site defense plan, and SOI to determine the specific requirements for the installation, operation, and maintenance of the signal center.	——	——
2. Implemented site defense plan.	——	——
3. Directed installation of secure RATT station. (Refer to applicable TMs.)	——	——
4. Directed installation of HF outstation radio net. (Refer to applicable TMs.)	——	——
5. Directed installation of telephone central office. (Refer to equipment TM.)	——	——
6. Directed installation of internal and external cable or field.	——	——
7. Directed installation of a telecommunications center.	——	——
8. Directed implementation of cover and concealment for personnel and equipment, as required. (Refer to FM 20-3.)	——	——
9. Coordinated with support personnel for POL, mess facilities, administrative needs of assigned personnel, maintenance, and field sanitation, as required.	——	——
10. Coordinated with the CE officer/noncommissioned officer for support of nonorganic communications requirements.	——	——

STP 31-18E34-SM-TG

Evaluation Guidance: Score the Soldier GO if all performance measures are passed (P). Score the Soldier NO-GO if any performance measure is failed (F). If the Soldier fails any performance measure, show what was done wrong and how to do it correctly.

References
    Required   **Related**
    FM 20-3

Subject Area 4: Communications Planning

Coordinate Signal Activities With Other Units
331-18E-4004

Conditions: Given the commander's planning guidance, communications-electronics (CE) annex to the operation order (OPORD), unit signal operating instructions (SOI) or extract, equipment lists, and pencil and paper.

Standards: Provide brief, specific, and timely instructions and a copy of the unit's SOI extract to all affected units.

Performance Steps

1. Determine the signal mission.
2. Determine how much and what type of communications equipment is needed and available.
3. Verify compatibility and frequency operating range of equipment.
4. Determine any auxiliary (AUX) signals (other than radio).
5. Forward any changes or additional information to be implemented into the CE annex of the OPORD to superiors for approval.
6. Review SOI or extract for completeness and clarity.
7. Brief and distribute the SOI or extract to all affected units.

Evaluation Preparation: Setup: Provide the Soldier with the commander's planning guidance, CE annex to the OPORD, unit SOI or extract, equipment lists, and pencil and paper.

Performance Measures	GO	NO-GO
1. Determined the signal mission.	——	——
2. Determined how much and what type of communications equipment is needed and available.	——	——
3. Verified compatibility and frequency operating range of equipment.	——	——
4. Determined any AUX signals (other than radio).	——	——
5. Forwarded any changes or additional information to be implemented into the CE annex of the OPORD to superiors for approval.	——	——
6. Reviewed SOI or extract for completeness and clarity.	——	——
7. Briefed and distributed the SOI or extract to all affected units.	——	——

Evaluation Guidance: Score the Soldier GO if all performance measures are passed (P). Score the Soldier NO-GO if any performance measure is failed (F). If the Soldier fails any performance measure, show what was done wrong and how to do it correctly.

STP 31-18E34-SM-TG

Supervise Signal Augmentation
331-18E-4005

Conditions: Given deployment instructions, operation order (OPORD), letter of instruction (LOI), or verbal orders of commanding officer (VOCO), list of personnel and equipment to become augmented, the communications plan, the signal operating instructions (SOI), and an operation site.

Standards: Verify that the deploying element is effectively augmented with signal support in accordance with commander's guidance, SOI, and all pertinent directives.

Performance Steps

1. Coordinate equipment and personnel requirements based upon OPORD, LOI, VOCO, or instructions.
2. Verify personnel security clearances with the intelligence officer (S-2).
3. Request radio frequencies.
4. Issue SOI or extracts of SOI to selected personnel.
5. Coordinate the establishment of the communications center (COMCEN) and related equipment (for example, antenna fields and power generation).
6. Select priority sites for telephone installation with the special operations task force.

Evaluation Preparation: Setup: Provide the Soldier with deployment instructions, OPORD, LOI, or VOCO, list of personnel and equipment to become augmented, the communications plan, the SOI, and an operation site.

Performance Measures	GO	NO-GO
1. Coordinated equipment and personnel requirements based upon OPORD, LOI, VOCO, or instructions.	——	——
2. Verified all security clearances with the S-2.	——	——
3. Requested the radio frequencies.	——	——
4. Issued SOI or extracts of the SOI to selected personnel.	——	——
5. Coordinated with COMCEN.	——	——
6. Selected priority sites for telephone installation.	——	——

Evaluation Guidance: Score the Soldier GO if all performance measures are passed (P). Score the Soldier NO-GO if any performance measure is failed (F). If the Soldier fails any performance measure, show what was done wrong and how to do it correctly.

References

Required	Related
	TC 24-21

STP 31-18E34-SM-TG

Glossary

Section I
Acronyms & Abbreviations

1SG	first sergeant
AC	active component; alternating current; hydrogen cyanide
ACCP	Army Correspondence Course Program
ACP	Allied Communications Publication
AD	armored division; air defense
ADA	air defense artillery
ADC	advanced data controller; area damage control
AIT	advanced individual training
altn	alternate
AM	amplitude modulation
amp	amperage; amplifier
AN	annually; artery to nerve
ANCD	automated net control device
ANCOC	advanced noncommissioned officer course
ANT	antenna
AO	area of operations; agent orange
app	appendix
AR	Army regulation; Army reserve; assistant rifleman; armor
ARTEP	Army Training and Evaluation Program
ASAT	Automated Systems Approach to Training
ASC	AUTODIN switching center
asst	assistance; assistant
ASUM	applications software user's manual
AT	annual training; antiterrorism; anti-tank
AUTODIN	Automatic Digital Network

AUX	auxiliary
BAT	battery
Bde	brigade
BIT	built-in-test
BLK DIG	black digital
BM	bimonthly
Bn	battalion
BNC	basic noncommissioned officer course (also BNCOC); connector that was named after its inventor (Bayonet Neill Concelman)
BNCOC	basic noncommissioned officer course
cav	cavalry
CBRN	Caribbean Basin Radar Network; chemical, biological, radiological, or nuclear
CBT	combatting terrorism; computer-based training
CD-ROM	compact disk-read-only memory
CE	communications-electronics
CEOI	communications-electronics operation instructions
CF	correlation factor; complement fixation
CH	chaplains
CIK	cryptographic ignition key
CMF	career management field
CO	company; commissioned officer; carbon monoxide; cardiac output
CO2	carbon dioxide
COMCEN	communications center
COMM	communications
COMSEC	communications security
CP	command post; counterproliferation; checkpoint
crypto	cryptographic
CSC	command and staff college

CSM	command sergeant major
DA	Department of the Army; direct action
DA Form	Department of the Army form
DA Pam	Department of the Army pamphlet
DC	dislocated civilian; Dental Corps; discharge; direct current; displaced civilian
DD	Defense Department
DISCOM	division support command
Div	division
DIVARTY	division artillery
DMD	digital message device
DOD	Department of Defense
DTS	data terminal software; Diplomatic Telecommunications Service
DZ	drop zone
EA (2)	electronic attack (previously ECM); emergency action; executive agent; executive assistant
ECCM	electronic counter-countermeasures
ECM	electronic countermeasures; erythema chronicum migrans
EEFI	essential elements of friendly information
ENL	erythema nodosum leprosum; enlisted
EP	electronic protection
EPLRS	enhanced position locating and reporting system
ESC	escape
EW	electronic warfare
FA	field artillery; functional area
FDC	fire direction center
FIST	fire support team
FM	field manual; frequency modulation
FRAGO	fragmentary order

STP 31-18E34-SM-TG

FSO	fire support officer
GPS	global positioning system
GSR	ground surveillance radar
GTA	graphic training aid
HF	high frequency
HHC	headquarters and headquarters company
I/O	input/output
IAW	in accordance with
ID	identification; intradermal
IN	interdiction; Infantry
INSCOM	U.S. Army Intelligence and Security Command
KTV	Cryptographic training, SOI
LDR	leader
Li-SO2	lithium-sulfur dioxide
LOC	lines of communications; location; level of consciousness
LOI	letter of instruction
LOS	line of sight
maint	maintenance
MBITR	multiband inter/intra team radio
MBMMR	multiband multimission radio
MED	medical
MEDEVAC	medical evacuation
MHz	megahertz
MIJI	meaconing intrusion jamming interference
MORT	mortar
MOS	military occupational specialty
MOSC	military occupational specialty code
MP	military police; mission planner

MTP	mission training plan; MOS training plan
NATO	North Atlantic Treaty Organization
NAVAIDS	navigational aids
NCO	noncommissioned officer
NCS	net control station; National Communications System
NO.	number
NSA	National Security Agency
NSN	national stock number
OCONUS	outside the continental United States
OPLAN	operation plan
OPORD	operation order
OPSEC	operations security
OS	operating system; outstation
Pam	pamphlet; pralidoxime
PC	personal computer
PDC	Psychological Operations development center; product development center; personal data controller
PLT	platoon
PMCS	preventive maintenance checks and services
POL	petroleum, oil, and lubricants
PRCD	portable remote-control device
PSA	primary support agency; port support activity; port shipping authority; power supply assembly
PT	physical training; prothrombin times; preparatory marksmanship training; plaintext; parachute team
PTT	push-to-talk
R/T	receiver/transmitter
RATT	radioteletype
RC	Reserve Component; reception committee

REF	reference
RF	radio frequency; reserve forces; rheumatoid factor
RFD	radio frequency direction
RMT	remote
RTO	radio telephone operator
S1	adjutant; first sacral vertebrae; first heart sound
S2	intelligence officer; second heart sound
S-2	intelligence officer
S3	operations and training officer; third heart sound
S-3	operations and training officer
S4	supply officer; fourth heart sound
sec	second
SF	Special Forces; standard form
SFOD	Special Forces operational detachment
SFODA	Special Forces operational detachment A
SFQC	Special Forces Qualification Course
SGM	sergeant major
SGT	sergeant
SIG	signal
SIGSEC	signal security
SIM	simulation; subscriber identity module
SINCGARS	single-channel ground and airborne radio system
SL	skill level; slight; squad leader
SM	Soldier's manual
SMCT	Soldier's manual of common tasks
SO2	sulfur dioxide
SOI	signal operating instructions
SOP	standing operating procedure

SORAK	special operations radio antenna kit
SOTF	special operations task force
SQD	squad
sqdn	squadron
ST	student text
STP	Soldier training publication
TACP	tactical air control party
TB	technical bulletin; tuberculosis
TC	technical coordinator; training circular
TG	task group; trainer's guide; training guide
TM	technical manual; tympanic membrane
TOC	tactical operations center
TOD	time of day
TOE	table of organization and equipment
TRANSEC	transportation security; transmission security
UHF	ultra-high frequency
US	United States
USAJFKSWCS	United States Army John F. Kennedy Special Warfare Center and School
V	nerve agent; volts; voice; volume; velocity
VAC	volts, alternating current
VDC	volts, direct current
VHF	very high frequency
VOCO	verbal orders of commanding officer
WIN	Windows
WPNS	weapons
XO	executive officer

Section II
Terms

Army Training and Evaluation Program (ARTEP)
The cornerstone of unit training. It is the umbrella program to be used by the trainer and training manager in the training evaluation of units. The ARTEP is a complete program enabling commanders to evaluate and develop collective training based on unit weaknesses, ten train the unit to overcome those weaknesses and reevaluate. Success on the battlefield depends on the coordinated performance of collective and individual skills that are taught through the ARTEP MTP.

common task
A critical task that is performed by every Soldier in a specific SL regardless of MOS.

cross-training
The systematic training of a Soldier on tasks related to another duty position within the same MOS or tasks related to a secondary MOS at the same SL.

Individual training
Training which prepares the Soldier to perform specified duties or tasks related to assigned duty position or subsequent duty positions and skill level.

merger training
Training that prepares NCOs to supervise one or more different MOSs at lower SLs when they advance to a higher SL in their career management field.

mission-essential task list
A compilation of collective mission-essential tasks that must be successfully performed if an organization is to accomplish its wartime mission(s).

Performance measures
The actions that can be objectively observed and measured to determine if a task performer has performed the task to the prescribed standard. These measures are derived from the task performance steps during task analysis. See "Task performance specifications."

train-up
The process of increasing the skills and knowledge of an individual to a higher SL in the appropriate MOS. It may involve certification.

unit training
Training (individual, collective, and joint or combined) that takes place outside the Army's institutional base.

STP 31-18E34-SM-TG

References

Other Product Types

ACP 125(F)	Communication Instructions—Radiotelephone Procedures, 5 September 2001
ACP 126(C)	Communication Instructions Teletypewriter (Teleprinter) Procedures, 10 May 1989
ACP 131(F)	Communications Instructions Operating Signals, 3 April 2006
Executive Order 12356	National Security Information, 2 April 1982

Manual for Courts Martial, United States, 1969 Edition (Revised)

Soldier Training Publications

STP 31-18-SM-TG	Soldier's Manual and Trainer's Guide, MOS 18, Special Forces Common Skills, Skill Levels 3 and 4, 24 October 2003

Technical Manuals

TM 11-5820-890-10-1	Operator's Manual for SINCGARS Ground Combat Net Radio, ICOM MANPACK Radio AN/PRC-119A (NSN 5820-01-267-9482) (EIC: L2Q) Short Range Vehicular Radio AN/VRC-87A (5820-01-267-9480) (EIC: L22) Short Range Vehicular Radio With Single Radio Mount AN/VRC-87C (5820-01-304-2045) (EIC: GDC) Short Range Vehicular Radio With Dismount AN/VRC-88A (5820-01-267-9481) (EIC: L23) Short Range Vehicular Radio With Dismount and Single Radio Mount AN/VRC-88C (5820-01-304-2044) (EIC: GDD) Short Range/Long Range Vehicular Radio AN/VRC-89A (5820-01-267-9479) (EIC: L24) Long Range Vehicular Radio AN/VRC-90A (5820-01-268-5105) (EIC: L25) Short Range/Long Range Vehicular Radio With Dismount AN/VRC-91A (5820-01-267-9478) (EIC: L26) Long Range/Long Range Vehicular Radio AN/VRC-92A (5820-01-267-9477) (EIC: L27), 1 September 1992
TM 11-5820-890-10-3	Operator's Manual for SINCGARS Ground Combat Net Radio, NON-ICOM MANPACK Radio AN/PRC-119 (NSN 5820-01-151-9915) (EIC: L2A) Short Range Vehicular Radio AN/VRC-87 (5820-01-151-9916) (EIC: L2T) Short Range Vehicular Radio (With Single Radio Mount) AN/VRC-87D (5820-01-351-5259) (EIC: TBD) Short Range Vehicular Radio With Dismount AN/VRC-88 (5820-01-151-9917) (EIC: L2U) Short Range Vehicular Radio With Dismount (With Single Radio Mount) AN/VRC-88D (5820-01-352-1694) (EIC: TDB) Short Range/Long Range Vehicular Radio AN/VRC-89 (5820-01-151-9918) (EIC: L2V) Long Range Vehicular RADIO AN/VRC-90 (5820-01-151-9919) (EIC: L2W) Short Range/Long Range Vehicular Radio With Dismount AN/VRC-91 (5820-01-151-9920) (EIC: L2X) Long Range/Long Range Vehicular Radio AN/VRC-92 (5820-01-151-9921) (EIC: L2Y), 1 September 1992
TM 11-5820-1130-12&P	Operator's and Unit Maintenance Manual (Including Repair Parts and Special Tools List) for Radio Set AN/PSC-5 (NSN 5820-01-366-4120) (EIC: N/A), 1 June 2000
TM 11-5985-357-13	Operator's, Organizational, and Direct Support Maintenance Manual for Antenna Group, OE-254/GRC (NSN 5985-01-063-1574), 1 February 1991

8 February 2010   References-1

STP 31-18E34-SM-TG

TM 11-5985-391-12	Operator's and Unit Maintenance Manual for Special Operations Radio Antenna Kit (SORAK) Antenna Group OE-452/PRC (NSN 5985-01-279-7942), 1 January 1991

Army Regulations
AR 25-2	Information Assurance, 24 October 2007
AR 380-5	Department of the Army Information Security Program, 29 September 2000
AR 380-40	(O) Policy for Safeguarding and Controlling Communications Security (COMSEC) Material, 30 June 2000
AR 530-1	(O) Operations Security (OPSEC), 19 April 2007

Department of Army Forms
DA FORM 2404	Equipment Inspection and Maintenance Worksheet
DA FORM 2407	Maintenance Request
DA FORM 2407-1	Maintenance Request Continuation Sheet
DA FORM 2653-R	COMSEC Account—Daily Shift Inventory
DA FORM 3964	Classified Document Accountability Record
DA FORM 5988-E	Equipment Inspection Maintenance Worksheet (EGA)
DA Forms are available on the APD web site (www.apd.army.mil).

Field Manuals
FM 3-13	Information Operations: Doctrine, Tactics, Techniques, and Procedures, 28 November 2003
FM 3-25.26	(O) Map Reading and Land Navigation, 18 January 2005
FM 5-424	Theater of Operations Electrical Systems, 25 June 1997
FM 6-02.53	Tactical Radio Operations, 5 August 2009
FM 20-3	Camouflage, Concealment, and Decoys, 30 August 1999
FM 24-1	Signal Support in the Airland Battle, 15 October 1990

Joint Publications
JP 3-13	Information Operations, 13 February 2006

Other Product Types
9505A PORTABLE PHONE	Iridium 9505A Portable Phone Quick Reference Guide
ACP 124(D)	Communication Instructions: Radiotelegraph Procedures, 1 October 1983
ISBN 0072122269	Networking: A Beginner's Guide by Bruce A. Hallberg, 22 December 1999
ISBN 0072337451	More Excellent HTML: With an Introduction to Javascript by Timothy T. Gottleber/Timothy N. Trainor, August 1999.
ISBN 0782122612	MCSE: Exchange Server 5.5 Study Guide by Richard Easlick and James Chellis, 17 April 1998
ISBN 0789710536	Upgrading and Repairing PCs by Scott Mueller, 1 December 1996
ISBN 1562057685	MCSE Training Guide: Windows NT Server and Workstation 4 by Joe Casad and Wayne Dalton, June 1997
ISBN 1562763644	How Computers Work by Ron White, 10 January 1995

Standard Forms
SF 153	COMSEC Material Report
SF 700	Security Container Information

STP 31-18E34-SM-TG

SF 702	Security Container Checksheet

Supply Bulletins

SB 11-131-2	Vehicular Radio Sets and Authorized Installations Volume II: SINCGARS, FHMUX, and EPLRS, 1 September 1985

Technical Bulletins

TB 11-5820-890-12	Operator and Unit Maintenance for AN/CYZ-10 Automated Net Control Device (ANCD) (NSN 5810-01-343-1194) (EIC: QSU) with the Single Channel Ground and Airborne Radio Systems (SINCGARS)(AR), 1 April 1993
TB 380-41	(O) Security Procedures for Safeguarding, Accounting and Supply Control of COMSEC Material, 15 March 2006

Technical Manuals

TM 11-2300-476-14&P	Operator's, Unit, Direct Support and General Support Maintenance Manual (Including Repair Parts and Special Tools List) for Installation Kits, Electronic Equipment: MK-2442/GRC-213 for Armored Personnel Carrier-M113A1 (NSN 5820-01-189-9791) MK-2443/GRC-213 for Utility Truck-M151A1 (5820-01-189-9792) MK-2444/GRC-213 for Cargo Truck-M561 (5820-01-189-9788) MK-2445/GRC-213 for Command Carrier Post-M577A1 (5820-01-189-9793) MK-2446/GRC-213 for Truck-M882 OR M1008A1 CUCV (5820-01-189-9789) MK-2447/GRC-213 for General Purpose Wheeled or Tracked Vehicles (5820-01-189-9790) MK-2542/GRC-213 for Cargo/Troop Carrier M998/M1038 HMMWV (5820-01-227-0528) MK-2543/GRC-213 for M1009 CUCV (5820-01-227-7088), 1 January 1987
TM 11-5810-410-13&P	(O) Operator's and Field Maintenance Manual Including Repair Parts and Special Tools List for Transfer Unit, Cryptographic Key AN/PYQ-10(C) Simple Key Loader (SKL) UAS Version 4.0 (NSN 5810-01-517-3587) (EIC: N/A), 1 September 2007
TM 11-5815-334-10	Operator's Manual for Radio Teletypewriter Sets, AN/GRC-122 (NSN 5815-00-401-9719) (EIC: GFE), AN/GRC-122A (5815-00-167-7998) (EIC: GFA), AN/GRC-122B (5815-00-937-5295) (EIC:GFJ), AN/GRC-122C (5815-01-095-1211) (EIC: GFL), AN/GRC-122D (5815-01-096-0428) (EIC: GFP), AN/GRC-122E(5815-01-095-1212) (EIC: GFM), AN/GRC-142 (5815-00-401-9720) (EIC: GFF), AN/GRC 142A (5815-00-168-1556) (EIC: GFB), AN/GRC-142B (5815-00-443-551) (EIC: GFG), AN/GRC-142C (5815- 01-100-6815) (EIC: GFR), AN/GRC-142D (5815-01-104-7264) (EIC: GFT), AN/GRC-142E (5815-01-095-6258) (EIC: GFN) Used with or without Installation Kit, Electronic Equipment Modification Kit MK-2488/G, 5 March 1985
TM 11-5820-467-15	Operator's, Organizational, Direct Support, General Support, and Depot Maintenance Manual for Antenna Group, AN/GRA-50 (NSN 5985-00-892-0758), 19 July 1961
TM 11-5820-890-10-8	Operator's Manual For SINCGARS Ground Combat Net Radio, ICOM MANPACK Radio, AN/PRC-119A (NSN 5820-01-267-9482) (EIC: L2Q), Short Range Vehicular Radio AN/VRC-87A (5820-01-267-9480) (EIC: L22), Short Range Vehicular Radio With Single Radio Mount AN/VRC-87C (5820-01-304-2045) (EIC: GDC), Short Range Vehicular Radio With Dismount AN/VRC-88A (5820-01-267-9481) (EIC: L23), Short Range/Long Range Vehicular Radio AN/VRC-89A (5820-01-267-9479) (EIC: L24), Long Range Vehicular Radio

STP 31-18E34-SM-TG

	AN/VRC-90A (5820-01-268-5105) (EIC: L25), Short Range/Long Range Vehicular Radio With Dismount AN/VRC-91A (5820-01-267-9478) (EIC: L26), Short Range/Long Range Vehicular Radio AN/VRC-92A (5820-01-267-9477) (EIC: L27) Used With Automated Net Control Device (ANCD) (AN/CYZ-10) Precision Lightweight GPS Receiver (PLGR) (AN/PSN-11) Secure Telephone Unit (STU) Frequency Hopping Multiplexer (FHMUX), 1 December 1998
TM 11-5820-890-20-1	Unit Maintenance Manual for Ground ICOM Radio Sets AN/PRC-119A (NSN 5820-01-267-9482) (EIC: L2Q) AN/PRC-119D (5820-01-421-0801) (EIC: N/A) AN/PRC-119F (5820-01-451-8252) (EIC: N/A) AN/VRC-87A (5820-01-267-9480) (EIC: L22) AN/VRC-87D (5820-01-351-5259) (EIC: GAR) AN/VRC-87F (5820-01-451-8248) (EIC: N/A) AN/VRC-88A (5820-01-267-9481) (EIC: L23) AN/VRC-88D (5820-01-352-1694) (EIC: GAS) AN/VRC-88F (5820-01-452-8435) (EIC: N/A) AN/VRC-89A (5820-01-267-9479) (EIC: L24) AN/VRC-89D (5820-01-420-6619) (EIC: N/A) AN/VRC-89F (5820-01-451-8247) (EIC: N/A) AN/VRC-90A (5820-01-267-5105) (EIC: L25) AN/VRC-90D (5820-01-420-6618) (EIC: N/A) AN/VRC-90F (5820-01-451-8246) (EIC: N/A) AN/VRC-91A (5820-01-267-9478) (EIC: L26) AN/VRC-91D (5820-01-420-6621) (EIC: N/A) AN/VRC-91F (5820-01-451-8249) (EIC: N/A) AN/VRC-92A (5820-01-267-9477) (EIC: L27) AN/VRC-92D (5820-01-421-2605) (EIC: N/A) AND AN/VRC-92F (5820-01-451-8250) (EIC: N/A), 30 December 1998
TM 11-5820-890-20-2	Unit Maintenance Manual for Ground ICOM Radio Sets AN/PRC-119A (NSN 5820-01-267-9482) (EIC: L2Q); AN/VRC-119D (5820-01-421-0801) (EIC: N/A); AN/VRC-87A (5820-01-267-9480) (EIC: L22); AN/VRC-87C (5820-01-304-2045) (EIC: GDC); AN/VRC-87D (5820-01-351-5259) (EIC: GAR); AN/VRC-88A (5820-01-267-9481) (EIC: L23); AN/VRC-88D (5820-01-352-1694) (EIC: GAS); AN/VRC-89A (5820-01-267-9479) (EIC: L24); AN/VRC-89D (5820-01-420-6619) (EIC: N/A); AN/VRC-90A (5820-01-267-5105) (EIC: L25); AN/VRC-90D (5820-01-420-6618) (EIC: N/A); AN/VRC-91A (5820-01-267-9478) (EIC: L26); AN/VRC-91D (5820-01-420-6621) (EIC: N/A); AN/VRC-92A (5820-01-267-9477) (EIC: L27); AN/VRC-92D (5820-01-421-2605) (EIC: N/A) (With Control, Receiver-Transmitter C-11561(C)/U (RCU)), 1 July 2000
TM 11-5820-924-13	Operator's, Organizational and Direct Support Maintenance Manual for Radio Set, AN/GRC-193A (NSN 5820-01-133-4195), 14 February 1986
TM 11-5830-340-12	Operator's and Unit Organizational Maintenance Manual For Intercommunication Set, AN/VIC-1(V) (NSN 5830-00-856-3273) and Control, Intercommunication Set, C-10456/VRC (5830-01-082-0804), 15 January 1986

Related Publications

Related publications are sources of additional information. They are not required in order to understand this publication.

Army Regulations
AR 220-1   Unit Status Reporting, 19 December 2006

Department of Army Pamphlets
DA PAM 750-8   The Army Maintenance Management System (TAMMS) Users Manual, 22 August 2005]

STP 31-18E34-SM-TG

Field Manuals
FM 3-19.30   Physical Security, 8 January 2001
FM 5-0   Army Planning and Orders Production, 20 January 2005
FM 11-65   High Frequency Radio Communications, 31 October 1978

Graphic Training Aids
GTA 31-01-003   Detachment Mission Planning Guide, 1 March 2006

Other Product Types
LSS-94429   LSS-94429 SOLSAT Phone Communication System Equipment Handbook

TT-98-107770B   TT-3060A Capsat Mobile Telephone Users Manual; Thrane & Thrane A/S, Denmark, 6 January 1998

Technical Manuals
TM 11-5820-520-10   Operator's Manual for Radio Sets, AN/GRC-106 (NSN 5820-00-402-2263) and AN/GRC-106A (5820-00-223-7548), 28 May 1984

TM 11-6625-203-12   Operation and Organizational Maintenance Manual for Multimeter AN/URM-105 and AN/URM-105C (Including Multimeters, ME-77/U and ME-77C/U), 11 June 1959

Training Circulars
TC 24-21   Tactical Multichannel Radio Communications Techniques, 3 October 1988

Department of Army Forms
DA FORM 2028   Recommended Changes to Publications and Blank Forms, 1 February 1974

DA Forms are available on the APD web site (www.apd.army.mil).

Soldier Training Publications
STP 21-1-SMCT   Soldier's Manual of Common Tasks, Warrior Skills, Level 1, 18 June 2009
STP 21-24-SMCT   Soldier's Manual of Common Tasks (SMCT), Warrior Leader Skill Level 2, 3, and 4, 9 September 2008

This page intentionally left blank.

STP 31-18E34-SM-TG
8 FEBRUARY 2010

By Order of the Secretary of the Army:

GEORGE W. CASEY, JR.
General, United States Army
Chief of Staff

Official:

JOYCE E. MORROW
Administrative Assistant to the
Secretary of the Army
1002801

DISTRIBUTION:

Active Army, Army National Guard, and United States Army Reserve: Not to be distributed. Electronic media only.

PIN: 080794-000